W9-BGM-716

NEWBURYPORT PUBLIC LIBRARY
94 STATE STREET
NEWBURYPORT, MA 01950

WITHDRAWN

CALCULATED
RISKS

HOW TO KNOW WHEN
NUMBERS DECEIVE YOU

Gerd Gigerenzer

SIMON & SCHUSTER

NEW YORK LONDON SYDNEY SINGAPORE TORONTO

SIMON & SCHUSTER
Rockefeller Center
1230 Avenue of the Americas
New York, NY 10020

Copyright © 2002 by Gerd Gigerenzer

All rights reserved,
including the right of reproduction
in whole or in part in any form.

SIMON & SCHUSTER and colophon are registered
trademarks of Simon & Schuster, Inc.

For information regarding special discounts for bulk purchases,
please contact Simon & Schuster Special Sales: 1-800-456-6798
or business@simonandschuster.com

Designed by Lisa Chovnick

Manufactured in the United States of America

10   9   8   7   6   5   4   3   2   1

Library of Congress Cataloging-in-Publication Data

Gigerenzer, Gerd.
Calculated risks : how to know when numbers deceive you / Gerd Gigerenzer.
p.   cm.
Includes bibliographical references and index.
1. Probabilities—Popular works.   2. Mathematical statistics—Popular works.
3. Uncertainty—Popular works   I. Title.

QA273.15 .G54 2002
519.2—dc21   2002017010

ISBN 0-7432-0556-1

For my mother

ACKNOWLEDGMENTS

Books, like human beings, have a history. They are conceived in love and written with perspiration. My fondness for the topic of uncertainty and risk was inspired by Ian Hacking's and Lorraine Daston's writings on the ideas of chance, rationality, and statistical thinking. David M. Eddy's work on medical diagnosis, and that of Lola L. Lopes on decision making and risk, have shown me how these ideas flourish and shape our present world. My interest in the way medical and legal experts think, and how one can offer them mind tools to better understand uncertainties, started with Ulrich Hoffrage, once my student and now my colleague and friend, whom I heartily thank for the fun of some ten years of joint research. This work has continued with Ralph Hertwig, Stephan Krauss, Steffi Kurzenhäuser, Sam Lindsey, Laura Martignon, and Peter Sedlmeier, as well as with other researchers at the Max Planck Institute for Human Development. Among the many people outside my research group who have shaped my thinking on the ideas presented in this book, I would like to thank Jonathan J. Koehler and John Monahan.

Many dear friends and colleagues have read, commented on, and helped shape the many versions of this book manuscript: Michael Birnbaum, Valerie M. Chase, Kurt Danziger, Norbert Donner-Banzhoff, George Daston, Robert M. Hamm, Ulrich Hoffrage, Max Houck, Günther Jonitz, Gary Klein, Jonathan J. Koehler, Hans-Joachim Koubenec, Steffi Kurzenhäuser, Lola L. Lopes, John Monahan, Ingrid Mühlhauser, Marianne Müller-Brettl, R. D. Nelson, Mike Redmayne, Joan Richards, Paul Slovic, Oliver Vitouch, William Zangwill, and Maria Zumbeel.

My special thanks go to Christine and Andy Thomson, one a lawyer and the other a psychiatrist, to whom I owe several of the case studies reported in this book.

Valerie M. Chase has edited the entire book manuscript and much clarity is due to her insight. Donna Alexander helped me at all stages of the development of the book, including the footnotes and references; she was a wonderful and critical support. Hannes Gerhardt joined in the final stages, Wiebke Möller helped find literature, even in the most remote places and Dagmar Fecht cleared the decks, making writing even possible.

Lorraine Daston, my beloved wife, gave me emotional and intellectual support during the four years I was gestating this book, and my daughter Thalia, always delightful and helpful, gave me valuable suggestions on improving the readability of the text.

People are important, but so is the environment in which one works. I have had the good fortune in the past few years to profit from the splendid intellectual atmosphere and the resources of the Max Planck Society, for which I am thankful.

CONTENTS

PART I

DARE TO KNOW

> ... in this world there is nothing certain but death and taxes.
>
> *Benjamin Franklin*

1

UNCERTAINTY

Susan's Nightmare

During a routine medical visit at a Virginia hospital in the mid-1990s, Susan, a 26-year-old single mother, was screened for HIV. She used illicit drugs, but not intravenously, and she did not consider herself at risk of having the virus. But a few weeks later the test came back positive—which at the time amounted to a terminal diagnosis. The news left Susan shocked and distraught. Word of her diagnosis spread, her colleagues refused to touch her phone for fear of contagion, and Susan eventually lost her job. Finally, she moved into a halfway house for HIV-infected patients. While there, she had unprotected sex with another resident, thinking, "Why take precautions if the virus is already inside of you?" Out of concern for her 7-year-old son's health, Susan decided to stop kissing him and began to worry about handling his food. The physical distance she kept from him, intended to be protective, caused her intense emotional suffering. Months later, she developed bronchitis, and the physician who treated her for it asked her to have her blood retested for HIV. "What's the point?" she thought.

The test came back negative. Susan's original blood sample was then retested and also showed a negative result. What had happened? At the time the data were entered into a computer in the Virginia hospital, Susan's original blood test result seems to have been inadvertently exchanged with those of a patient who was HIV positive. The error not only gave Susan false despair, but it gave the other patient false hope.

The fact that an HIV test could give a false positive result was news to Susan. At no point did a health care provider inform her that laboratories, which run two tests for HIV (the ELISA and Western blot) on each blood sample, occasionally make mistakes. Instead, she was told repeatedly that HIV test results are absolutely conclusive—or rather, that although one test might give false positives, if her other, "confirmatory" test on her initial blood sample also came out positive, the diagnosis was absolutely certain.

By the end of her ordeal, Susan had lived for 9 months in the grip of a terminal diagnosis for no reason except that her medical counselors believed wrongly that HIV tests are infallible. She eventually filed suit against her doctors for making her suffer from the illusion of certainty. The result was a generous settlement, with which she bought a house. She also stopped taking drugs and experienced a religious conversion. The nightmare had changed her life.

Prozac's Side Effects

A psychiatrist friend of mine prescribes Prozac to his depressive patients. Like many drugs, Prozac has side effects. My friend used to inform each patient that he or she had a 30 to 50 percent chance of developing a sexual problem, such as impotence or loss of sexual interest, from taking the medication. Hearing this, many of his patients became concerned and anxious. But they did not ask further questions, which had always surprised him. After learning about the ideas presented in this book, he changed his method of communicating risks. He now tells patients that out of every ten people to whom he prescribes Prozac, three to five experience a sexual problem. Mathematically, these numbers are the same as the percentages he used before. Psychologically, however, they made a difference. Patients who were informed about the risk of side effects in terms of frequencies rather than percentages were less anxious about taking Prozac—and they asked questions such as what to do if they were among the three to five people. Only then did the psychiatrist realize that he had never checked how his patients understood what "a 30 to 50 percent chance of developing a sexual problem" meant. It turned out that many of them had thought that something would go awry in 30 to 50 percent of their sexual encounters. For years, my friend had simply not noticed that what he intended to say was not what his patients heard.

The First Mammogram

When women turn 40, their gynecologists typically remind them that it is time to undergo biennial mammography screening. Think of a family friend of yours who has no symptoms or family history of breast cancer. On her physician's advice, she has her first mammogram. It is positive. You are now talking to your friend, who is in tears and wondering what a positive result means. Is it absolutely certain that she has breast cancer, or is the chance 99 percent, 95 percent, 90 percent, 50 percent, or something else?

I will give you the information relevant to answering this question, and I will do it in two different ways. First I will present the information in probabilities, as is usual in medical texts.[1] Don't worry if you're confused; many, if not most, people are. That's the point of the demonstration. Then I will give you the same information in a form that turns your confusion into insight. Ready?

> *The probability that a woman of age 40 has breast cancer is about 1 percent. If she has breast cancer, the probability that she tests positive on a screening mammogram is 90 percent. If she does not have breast cancer, the probability that she nevertheless tests positive is 9 percent. What are the chances that a woman who tests positive actually has breast cancer?*

Most likely, the way to an answer seems foggy to you. Just let the fog sit there for a moment and feel the confusion. Many people in your situation think that the probability of your friend's having breast cancer, given that she has a positive mammogram, is about 90 percent. But they are not sure; they don't really understand what to do with the percentages. Now I will give you the same information again, this time not in probabilities but in what I call *natural frequencies:*

> *Think of 100 women. One has breast cancer, and she will probably test positive. Of the 99 who do not have breast cancer, 9 will also test positive. Thus, a total of 10 women will test positive. How many of those who test positive actually have breast cancer?*

Now it is easy to see that only 1 woman out of 10 who test positive actually has breast cancer. This is a chance of 10 percent, not 90 percent. The fog in your mind should have lifted by now. A positive mammogram is not good news. But given the relevant information in natural frequencies, one can see that the majority of women who test positive in screening do not really have breast cancer.

DNA Tests

Imagine you have been accused of committing a murder and are standing before the court. There is only one piece of evidence against you, but it is a potentially damning one: Your DNA matches a trace found on the victim. What does this match imply? The court calls an expert witness who gives this testimony:

> "The probability that this match has occurred by chance is 1 in 100,000."

You can already see yourself behind bars. However, imagine that the expert had phrased the same information differently:

> "Out of every 100,000 people, 1 will show a match."

Now this makes us ask, how many people are there who could have committed this murder? If you live in a city with 1 million adult inhabitants, then there should be 10 inhabitants whose DNA would match the sample on the victim. On its own, this fact seems very unlikely to land you behind bars.

Technology Needs Psychology

Susan's ordeal illustrates the *illusion of certainty;* the Prozac and DNA stories are about *risk communication;* and the mammogram scenario is about *drawing conclusions* from numbers. This book presents tools to help people to deal with these kinds of situations, that is, to understand and communicate uncertainties.

One simple tool is what I call "Franklin's law": *Nothing is certain but death and taxes.*[2] If Susan (or her doctors) had learned this law in school, she might have asked immediately for a second HIV test on a different blood sample, which most likely would have spared her the nightmare of living with a diagnosis of HIV. However, this is not to say that the results of a second test would have been absolutely certain either. Because the error was due to the accidental confusion of two test results, a second test would most likely have revealed it, as later happened. If the error, instead, had been due to antibodies that mimic HIV antibodies in her blood, then the second test might have confirmed the first one. But whatever the risk of error, it was her doctor's responsibility to inform her that the test results were uncertain. Sadly, Susan's case is not an exception. In this book, we will meet medical experts, legal experts, and other professionals who continue to tell the lay public that DNA fingerprinting, HIV tests, and other modern technologies are foolproof—period.

Franklin's law helps us to overcome the illusion of certainty by making us aware that we live in a twilight of uncertainty, but it does not tell us how to go one step further and deal with risk. Such a step is illustrated, however, in the Prozac story, where a *mind tool* is suggested that can help people understand risks: *When thinking and talking about risks, use frequencies rather than probabilities.* Frequencies can facilitate risk communication for several reasons, as we will see. The psychiatrist's statement "You have a 30 to 50 percent chance of developing a sexual problem" left the *reference class* unclear: Does the percentage refer to a class of people such as patients who take Prozac, to a class of events such as a given person's sexual encounters, or to some other class? To the psychiatrist it was clear that the statement referred to his patients who take Prozac, whereas his patients thought that the statement referred to their own sexual encounters. Each person chose a reference class based on his or her own perspective. Frequencies, such as "3 out of 10 patients," in contrast, make the reference class clear, reducing the possibility of miscommunication.

My agenda is to present mind tools that can help my fellow human beings to improve their understanding of the myriad uncertainties in our modern technological world. The best technology is of little value if people do not comprehend it.

Dare to know!

Kant

2

THE ILLUSION OF CERTAINTY

The creation of certainty seems to be a fundamental tendency of human minds.[1] The perception of simple visual objects reflects this tendency. At an unconscious level, our perceptual systems automatically transform uncertainty into certainty, as depth ambiguities and depth illusions illustrate. The Necker cube, shown in Figure 2-1, has ambiguous depth because its two-dimensional lines do not indicate which face is in front and which is in back. When you look at it, however, you do not see an ambiguous figure;

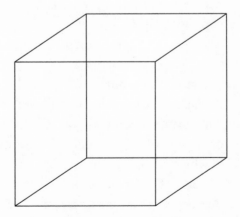

FIGURE 2-1. *The Necker cube.* If you fixate on the drawing, your perceptual impression shifts between two different cubes—one projecting into and the other out of the page.

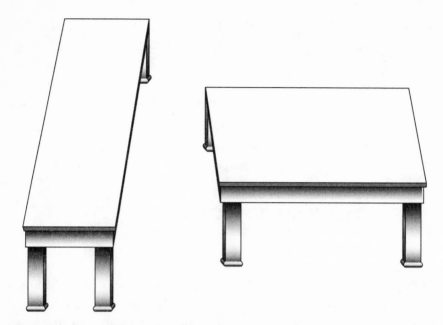

FIGURE 2-2. *Turning the tables.* These two tables are of identical size and shape. This illusion was designed by Roger Shepard (1990). (Reproduced with permission of W. H. Freeman and Company.)

you see it one way or the other. After a few seconds of fixating on the cube, however, you experience a gestalt switch—that is, you see the other cube instead, but again unambiguously.

Roger Shepard's "Turning the Tables," a depth illusion shown in Figure 2-2, illustrates how our perceptual system constructs a single, certain impression from uncertain cues. You probably see the table on the left as having a more elongated shape than the one on the right. The two surfaces, however, have exactly the same shape and area, which you can verify by tracing the outlines on a piece of paper. I once showed this illustration in a presentation during which I hoped to make an audience of physicians question their sense of certainty ("often wrong but never in doubt"). One physician simply did not believe that the areas were the same shape. I asked him how much he wanted to bet, and he offered me $250. By the end of my talk, he had disappeared.

What is going on in our minds? Unconsciously, the human perceptual system constructs representations of three-dimensional objects from incomplete information, in this case from a two-dimensional drawing. Consider the longer sides of each of the two tables. Their projections on the retina have the same length. But the perspective cues in the drawings indicate that the longer side of the left-hand table extends into depth, whereas that of the right-hand table does not (and vice versa for their shorter sides). Our perceptual systems assume that a line of a given length on the retina that extends into depth is actually longer than one that does not and corrects for that. This correction makes the left-hand table surface appear longer and narrower.

Note that the perceptual system does not fall prey to illusory certainty—our conscious experience does. The perceptual system analyzes incomplete and ambiguous information and "sells" its best guess to conscious experience as a definite product. Inferences about depth, orientation, and length are provided automatically by underlying neural machinery, which means that any understanding we gain about the nature of the illusion is virtually powerless to overcome the illusion itself. Look back at the two tables; they will still appear to be different shapes. Even if one understands what is happening, the unconscious continues to deliver the same perception to the conscious mind. The great nineteenth-century scientist Hermann von Helmholtz coined the term "unconscious inference" to refer to the inferential nature of perception.[2] The illusion of certainty is already manifest in our most elementary perceptual experiences of size and shape. Direct perceptual experience, however, is not the only kind of belief where certainty is manufactured.

Technology and Certainty

Fingerprinting has enjoyed an impeccable reputation. The fingerprints themselves are unique features of an individual and persist unchanged from early youth into advanced age. Even identical twins have different fingerprints despite having the same genes. If the fingerprints of a suspect matched those found at the scene of the crime, what jury would acquit the suspect? Fingerprint evidence seems absolutely certain—*the* great exception to Franklin's law, it would seem.

The use of fingerprinting acquired a scientific basis through the work of Sir Francis Galton, a nineteenth-century English gentleman scientist and Charles Darwin's first cousin. Galton looked at the arches, whorls, and loops of which fingerprints consist and estimated the chance that two fingerprints would match randomly as 1 in 64 billion.[3] Galton did not look at the whole fingerprint, but at points ("points of similarity") where the ridges in the fingerprint either end or split. His estimate was based on using every point, and there are typically 35 to 50 such points in a fingerprint. However, current practice has been to declare a match between the fingerprints of a suspect and those found at a crime scene when 8 to 16 of these points match. (This practice varies widely.) But in England an alternative technique is being implemented that relies on an examiner's overall impression of a match, not on counting the matches between points. This alternative makes the determination of a match a subjective judgment. The validity of neither of the two techniques—determining points of similarity and overall impression—however, has been scientifically investigated. Fingerprint experts have few statistics on which they can base their conclusions.

When actual fingerprints are found at a crime scene, two complicating factors enter: fingerprints are typically incomplete, and they are "latent." With incomplete fingerprints, one cannot compare whole prints, only fragments. Galton's statistical analysis and its modern equivalents do not help much in this case. The second complication is that most fingerprints found at the scene are latent; that is, they require treatment with chemicals or illumination with ultraviolet light to make them visible enough to work with. How reliable is it to compare such filtered evidence with the suspect's clean fingerprints taken under controlled conditions? Given these uncertainties and differences in procedure, how certain is fingerprint evidence? The answer is that we do not know; there seem to be no reliable scientific studies.

Recently, however, the Federal Bureau of Investigation performed a test of the reliability of fingerprint evidence that had never been done before. In 1998, Byron Mitchell appealed his conviction for having driven the getaway car in a robbery in Pennsylvania in 1991. The conviction was based on two latent fingerprints, one found on the steering wheel and the other on the gearshift of the car. The FBI decided to test the reliability of the reported match and sent the latent fingerprints along with Mr. Mitchell's inked prints to the laboratories of 53 various state law enforcement agencies. From the 35 laboratories that responded, 8 failed to find a match for one of the prints, and 6 failed to find a match for the other, making the average failure rate one in five tests. This troublesome result casts considerable doubt on the reliability of fingerprinting. America's National Institute of Justice has finally provided funding to study how good fingerprinting actually is.[4]

Fingerprint evidence has been accepted as certain for more than a century, following Galton's estimate. This calculation was made for ideal conditions, which are not found in the real world of incomplete and latent prints. When DNA fingerprinting was introduced into the courts, almost a hundred years after Galton's seminal work, the public and many experts projected the illusion of certainty onto this new technology. As we see in Chapter 10, DNA fingerprinting has also been declared "failsafe." The fiction that fingerprinting, DNA fingerprinting, HIV tests, or other excellent new technologies are absolutely foolproof is like a dream that comes back night after night, fulfilling a deep unconscious wish.

Authority and Certainty

When I was a child, I was told on good authority never to drink water after eating cherries, or I would get very sick and might even die. It never occurred to me to doubt this warning. One day I shared an ample serving of cherries with an English friend who had never heard of this danger. To my horror, I saw him reach for a glass of water after eating some of the cherries. I tried to stop him, but without success; he just laughed. He took a sip, and nothing happened. Not only did he not die; he did not even get sick. That experience cured me.[5] My belief about cherries was groundless, as many convictions about food and dieting are. However, this is not always the case. The general disposition to uncritically accept such a belief is not unfounded when beliefs involve food, health, or other things that directly concern survival. Here, illusory certainty seems to be an adaptive response that for ages has protected humans, especially children, from trying to learn first-hand about possible dangers, such as which kinds of food are poisonous and which are not. Similarly, young children are prepared to believe in values, rules, and stories without question, which facilitates their integration into their social group and culture. Social conventions—whether learned from one's family or one's wider culture—are, like elementary perception, a source of the illusion of certainty.

Illusory certainty is part of our perceptual, emotional, and cultural inheritance. It can provide us with images of our environment that are useful, although not always correct, as well as with feelings of comfort and safety. The esoterica sections in today's bookstores attest to the idea that many people crave fast faith. Throughout history, humans have created belief systems that promise certainty, such as religion, astrology, and divination, systems in which people—particularly those experiencing terrible suffering—can find comfort. Certainty has become a consumer product. It is marketed the world over—by insurance companies, investment advisers, election campaigns, and the medical industry. In seventeenth-century Europe, buying life insurance meant making a bet on the duration of a prominent person's life, such as whether the mayor of Paris would die within three years. If he died within the period on which you had wagered, you made a small fortune. Nowadays, insurance agencies have persuaded us that life insurance is about safety and certainty and that it is morally responsible to bet against our own lives for the sake of our survivors' economic security.[6] Political parties likewise fuel the desire for security. Before Germany's national elections of 1998, streets all over the country were plastered with Christian Democratic campaign posters that read: "CERTAINTY, NOT RISK." This promise was not unique to the platform of Chancellor Helmut Kohl and his party; other parties in the elections also advertised certainty.

The illusion of certainty can be created and exploited as a tool for political and economic goals. In recent years, for instance, when mad cow disease (bovine spongiform encephalopathy, or BSE) raged in Great Britain, Ireland, Portugal, France, and Switzerland, the German government declared its country BSE-free. "German beef is safe"—this phrase was repeated again and again by the president of the Farmers' Association, the minister of agriculture, and a choir of other government officials.[7] The German people liked to hear this message. English beef had been banned, and customers were advised to ask their butcher for beef bred in Germany. In other countries, so they were told, there was a marked lack of care and control.

When, in the year 2000, the Germans finally began to actually perform a substantial number of tests on their own herds for BSE, the disease was found, and the public was caught totally by surprise. Ministers were forced to resign, beef prices toppled, and other countries banned German beef. The government finally admitted that it had clung too long to the illusion that German cattle were entirely free of the disease.

However, the game of promising certainty did not stop; only the players changed. Supermarkets and butchers now put up signs and set out pamphlets reassuring their customers: "Our beef is guaranteed BSE-free." Some explained that this was because their cows had, luckily, grazed on ecological meadows, and others that their cows had actually been tested—none mentioned that these tests contain a large number of errors. When the newspapers finally reported the case of a cow that tested negative despite, indeed, having BSE, the public was again given a shock. Yet another illusion of certainty disappeared. Reassurance rather than information concerning BSE was the primary goal of both the government and the butchers and supermarkets.

Political and marketing campaigns show that illusory certainty does not begin or end with the individual mind. Many parties may be involved in the creation and selling of certainty, such as members of a profession who publicly deny the possibility that their products could be flawed, and clients who want to hear and trust this message and surrender to social authority. The illusion is also not intended for or created in all minds; it may be conjured up for specific audiences. For instance, the Yale Law School professor Dr. Jay Katz recounted a discussion he had had with a surgeon friend about the uncertainties that plague breast cancer treatment, during which they had agreed that nobody knows what the single best treatment is.[8] When Katz asked his friend how he advised his patients, the surgeon recounted telling his most recent patient with breast cancer that the single best treatment is radical surgery and impressing upon her the need to have the operation. Katz challenged his friend about this inconsistency: How could he suddenly be so sure what the best course of action was? Although he admitted that he hardly knew the patient, the surgeon insisted that his patients—this woman included—would neither comprehend nor tolerate knowledge of the uncertainties inherent in choosing treatment. In his view, patients want the illusion of certainty, and this patient got it.

In what follows, I invite you to have a closer look at the tangled web of motives from which the illusion of certainty is fabricated. We will take a look at the physician-patient relationship from the inside, from the point of view of physicians who discuss the pros and cons of the illusion of certainty.

Physicians on Certainty and Responsibility

In 2000, I attended a meeting of 60 physicians, including representatives of physicians' organizations and health insurance companies. All of them were interested in *evidence-based medicine,* in which physician and patient base medical decisions on the available evidence rather than on mere conviction, preference, or habit. These people came from several countries in Europe and the United States to a scenic vacation spot to spend two days together. The topic of the meeting was how to improve risk communication, physician-patient interaction, and public knowledge about medical screening. The atmosphere was casual. The organizer's warm personality and the beautiful setting helped us to develop a sense of trust and a common agenda. On the second day, an intense discussion took place on physicians' responsibility and patients' illusory certainty. Here is an exchange from the middle of this discussion:[9]

Dr. A: We doctors are the victims of the images we have of our patients: we think the patient is not capable of informing himself.

Representative of the World Health Organization (WHO): In the U.S., the average physician-patient contact is five minutes. Most of the information is presented in a vocabulary that is unintelligible to the patient. Patients tend to develop views of "fate" or "Inshallah" rather than learning to practice informed consent. All is God's will, or the physician's; why should they worry? The Institute of Medicine estimated that some 44,000 to 98,000 patients are killed every year in U.S. hospitals by preventable medical errors and misadventures. It's as if one lived in a culture where death is a desirable transition from one life to a better one.

Dr. B: Isn't that a bit much? That's saying more people die from hospital accidents than from motor vehicle accidents, or from AIDS.

WHO: It's based on the records of hospitals in New York, Colorado, and Utah. These errors were preventable, such as when a physician prescribed an antibiotic to a patient with a history of documented allergic reactions without consulting the medical records. The general problem in medicine is that, unlike in aviation, there is no system for reporting errors without punishing the individual doctor. Pilots anonymously report "near misses" to a central database so that other pilots can learn from them and improve air transport safety. Aviation has focused on building safe systems since World War II, and U.S. airline fatalities have decreased ever since. In 1998, there were no deaths in the United States in commercial aviation. Health care has no such system.

Dr. A: Women go to screening to be sure that they do not have cancer. But mammograms don't deliver certainty; they miss some 10 percent of cancers. And screening has both possible benefits and harms, of which most women are not informed. They just don't know.

Dr. B (after murmuring skeptically):   Informed consent—that's just a politically correct tale. If I were to start explaining to patients the benefits and harms of a potential treatment, they would hardly comprehend it. And if I were, in addition, to tell them what we do not know, they would get very nervous.

Dr. C: I agree. Sixty percent of patients, conservatively estimated, do not have the intellectual capacity to make decisions about treatments themselves.

Breast cancer specialist: Let's talk about physicians, not patients. The major source of continuing education for physicians is seminars run by

pharmaceutical firms. The best hotel in town, opulent dinners, partners invited. When we offer seminars for continuing education, with a
more unbiased view of the matter, all my institute can afford is a dull
lecture room, and there is no money for drinks or even fast food. We
can attract few physicians. Concerning the intellectual capacities of
patients, I have decided to discuss with patients the pros and cons of
hormone therapy—such as less depression on the one side and higher
risk, a factor of 1.4, of breast cancer on the other. The problem is not
that women have too low an IQ to make their own decisions. They do
make them if you provide the information. My problem is that since
my patients have started making their own decisions, my colleagues
send me fewer of their patients.

Dr. A: When our organizer gave a talk at our institute, women left, un-
nerved, in droves. Some went to natural healers, back to the promise
of certainty.

Dr. B: But how could I allow a patient to decide for herself? How can one
be a responsible physician and still allow the patient to make the decision?

Professor O: Look, I have two sons; both attend school. In that school the
rule is if pupils want to go home because they feel sick, they are sent to
a school physician. This man routinely X-rays every child; that is, if a
boy complains that his hand hurts, his hand is X-rayed; if his chest
hurts, his chest is X-rayed. Just to be on the safe side. Typically, the
physician concludes that there is no fracture, only serious bruising.
Some children just want to cut class so they pretend to be in pain and
are X-rayed. I told my boys that they are not allowed to have these
X rays, and that they should tell the doctor that their dad is a physician. As a father, I have the responsibility for my children, and I should
not delegate this to the next physician.

Dr. B: I think all this talk about informed consent—benefits and costs—
misses the point. The meeting between physician and patient is a ritual. False positives have no place in this ritual.

Several physicians (agitated):   That's right, rituals. That's what it is all
about.

President of a medical association: Patients want to be reassured. They
want to be relieved of their anxiety, to be in the right hands, even if

they do not feel better than before. They want a label for their suffering. A physician who takes anxiety away from the patient is a good doctor. One has to do something; one cannot do nothing; the patient would be disappointed or even angry. Most prescriptions have no proven effect, but when the patient applies the ointment, the doctor, the patient, and the pharmaceutical company are happy.

Radiologist: It is not money that drives physicians—it is salvation. The physician as a hero. Heroism is self-deception and the greatest obstacle towards progress.

President: If the doctor explains to the patient the risk in terms of "number needed to treat," then the placebo effect is gone. After all, "number needed to treat" means how many people have to suffer treatment so that just one can be saved. A patient visits a doctor to be cured, not to learn how many have to be harmed so that one can be cured.

Dr. C: When it comes to health, rituals are unavoidable. From an economic point of view, screening often does not pay; the tax dollars could be spent on other things that are more beneficial. But for the physician-patient relationship, it pays.

President: The areas where medicine has made real progress quickly disappear from public attention. Everything concentrates on areas and treatments of questionable value. There are too many doctors, too little money, and false incentives—a situation that reminds one of rats crowded together that do strange things. And there is the ideal of infallibility. Patients want to believe in a doctor who never errs, and doctors try to foster this illusion.

WHO: Uncertainty is a threat to practitioners. It's hard to say "I don't know."

This discussion reveals physicians' complex motives, emotions, and beliefs about the illusion of certainty, the existence of which none of the discussants denied. Different physicians struggle with this conundrum in different ways. Should a physician destroy patients' illusions and reveal to them the uncertainties involved in a treatment? Should a physician always say "I don't know" when she or he doesn't know?

One group of physicians at this meeting was convinced that uncertain-

ties should not be fully disclosed. Some believed that most patients would
not understand the uncertainties in the first place, would become more
confused upon learning of them, and might even decide to go to another
healer, one who offers certainty. Others hinted at pragmatic constraints; it
is hard to fully inform patients about risks during an interaction that lasts
an average of five minutes. In these physicians' view, the patient wants to be
reassured, not informed; they see this interaction as a ritual for generating
the feeling that the patient is being taken care of.

A second group of physicians, in contrast, felt strongly that authority
and emotional reassurance are not all that patients need and that many patients are capable of dealing with uncertainty. The breast cancer specialist,
for instance, had little respect for his colleagues' patronizing attitude
toward the patient as an intellectual inferior. He pointed out that there is
ignorance on both sides of the physician-patient relationship and that informed patients are not always welcome in physicians' offices. In his experience, informing patients meant losing referrals.

The president made the challenging argument that neither reassurance
nor informed consent is the best choice in every situation. He used the
placebo effect to illustrate his point. The *placebo effect* is a well-known phenomenon in medicine and psychotherapy in which a treatment tends to
have at least some positive effect if patients *believe* it is beneficial. One explanation for this effect is that a patient's belief mobilizes forces in the immune system that have been kept in reserve, much as a distance runner can
mobilize all his reserves the moment he realizes that the finish line is near.[10]
The president called attention to the phenomenon that the placebo effect
might disappear as soon as the physician explained to patients the actual
risks inherent in the treatment, that is, as soon as the physician assumed a
role founded on reason rather than authority. If illusions can sometimes
cure, then there is a problem—the benefit of knowing is not absolute; there
is also the potential efficacy of faith.[11]

This discussion also highlights the possibility that physicians and patients can have different or even opposing goals. The school physician routinely performs X rays; his goal is to protect himself from possible
accusations of overlooking a fracture. However, Professor O's goal is to
protect his sons from being harmed by the X rays. Each of the two alterna-

tives—to X-ray or not—involves potential benefits and costs for his boys, but these are not the same as those for the school physician. The breast cancer expert informs patients about the pros and cons of hormone therapy so that each woman can make her own decision, depending on what is more important to her. For instance, a woman may prefer lowering her chance of depression over reducing her chance of developing breast cancer, or vice versa. The expert's tutorial is in the women's interest, but not necessarily in that of his fellow physicians who definitely want women to *receive* hormone therapy, not to reflect about it. The representative of WHO reported the shocking number of patients killed every year in U.S. hospitals by preventable errors, but safety systems such as in commercial aviation that would be in the interest of patients have not been set up in hospitals. In aviation, safety is in the immediate interest of the pilot; if the passengers die in a crash, the pilot will very likely die too. The situation for the patient vis-à-vis the doctor is different.

Because costs and benefits typically differ for physicians and patients, it is imperative that patients be informed and be in a position to choose their treatment on this basis. Patients' choices will not be, and should not always be, the same as their physician's, and a good doctor will reveal to the patient when their interests diverge. The illusion of certainty—such as that treatments have only benefits but not any harm; that there is one and only one best treatment; that a diagnostic test is absolutely certain—is a mental obstacle toward making up one's own mind.

Kant's Dream

In his essay "What Is Enlightenment?" the philosopher Immanuel Kant begins thus:

> Enlightenment is man's emergence from his self-imposed nonage. Nonage is the inability to use one's own understanding without another's guidance. This nonage is self-imposed if its cause lies not in lack of understanding but in indecision and lack of courage to use one's own mind without another's guidance. Dare to know![12]

These are lucid and lovely sentiments. The key term is "courage." Courage is necessary because using one's own mind can bring not only feelings of liberation and autonomy, but also punishment and pain. Kant himself had to experience this. A few years after he wrote these lines, he was required by the government—out of fear that his rational thinking would undermine the certainty of the Christian doctrine—to cease writing and lecturing on religious subjects. In general, overcoming nonage can mean detecting holes in stories, facts, and values in which one has always believed. Questioning certainties often means questioning social authority.

Learning to live with uncertainty is a daring task for individuals as well as societies. Much of human history has been shaped by people who were absolutely certain that their kin, race, or religion was the one most valued by God or destiny, which made them believe they were entitled to get rid of conflicting ideas along with the bodies polluted with them. Modern societies have come a long way toward greater tolerance of uncertainty and diversity. Nevertheless, we are still far from being the courageous and informed citizens whom Kant envisaged—a goal that can be expressed in just two Latin words: *Sapere aude.* Or in three English words: "Dare to know."

Math is hard. Let's go shopping!

Barbie[1]

3

INNUMERACY

At the beginning of the twentieth-century, the father of modern science fiction, H. G. Wells, is reported to have predicted, "Statistical thinking will one day be as necessary for efficient citizenship as the ability to read and write."[2] At the end of the century, the mathematician John Allen Paulos investigated how far we had—or, rather, hadn't—come in this respect. In his best-selling book, *Innumeracy,* Paulos related the story of a weather forecaster on American television who reported that there was a 50 percent chance of rain on Saturday and a 50 percent chance of rain on Sunday, from which he concluded that there was a 100 percent chance of rain that weekend!

The inability to reason appropriately about uncertainties is by no means strictly an American affliction. The word "percentage" has become one of the most frequent nouns in the German media. In a survey, 1,000 Germans were asked what "40 percent" means: *(a)* one-quarter, *(b)* 4 out of 10, or *(c)* every 40th person. About one-third of respondents did not choose the right answer.[3] Political decision makers are, likewise, not immune to innumeracy. For example, commenting on the dangers of drug abuse, a Bavarian minister of the interior once argued that because most heroin addicts have used marijuana, most marijuana users will become heroin addicts. Figure 3-1 shows why this conclusion is mistaken. Most heroin addicts indeed have used marijuana, as the dark section of the small circle shows. However, this does not mean that most marijuana users are heroin addicts—the same dark section that covers most of the heroin addicts covers only a small portion of the marijuana users. On the basis of his mistaken conclusion, the minister of the interior asserted that marijuana should therefore remain illegal. Whatever one's views on the legalization of marijuana, the minister's conclusion was based on clouded thinking.

In Western countries, most children learn to read and write, but even in adulthood, many people do not know how to think with numbers. This is the problem that Paulos and others have called *innumeracy.* I focus on the most important form of innumeracy in everyday life, statistical innumeracy—that is, the inability to reason about uncertainties and risk. Henceforth, when I use the term "innumeracy," I mean statistical innumeracy. How is the illusion of certainty connected to innumeracy? Here is an overview.

- *Illusion of certainty.* Franklin's law is a mind tool to overcome the illusion of certainty, to help make the transition from certainty to uncertainty. For instance, when Susan, the woman introduced in Chapter 1, finally learned (the hard way) that labora-

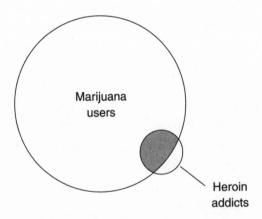

FIGURE 3-1. *Marijuana users and heroin addicts.* Most heroin addicts are marijuana users (the dark part of the smaller circle). Can we conclude from this that most marijuana users are heroin addicts?

tory errors occur in HIV testing, she made the transition from certainty to uncertainty.

- *Ignorance of risk.* This is an elementary form of innumeracy in which a person does not know, not even roughly, how large a personally or professionally relevant risk is. This differs from the illusion of certainty in that the person is aware that there may be uncertainties, but does not know how great these are. The major tool for overcoming the ignorance of risk consists of various forms of information search (for example, scientific literature). For instance, Chapter 7 gives details about the various risks involved in HIV testing, including false positives.

- *Miscommunication of risk.* In this form of innumeracy a person knows the risks but does not know how to communicate these so that others understand them. The mind tool for overcoming miscommunication is representations that facilitate understanding. For instance, the Prozac story in Chapter 1 illustrates the miscommunication of risk—the failure to communicate risk in an understandable way—and how to overcome it.

- *Clouded thinking.* In this form of innumeracy a person knows the risks but not how to draw conclusions or inferences from them. For instance, physicians often know the error rates of a clinical test and the base rate of a disease, but not how to infer from this information the chances that a patient with a positive test actually has the disease (Chapter 1). Representations such as natural frequencies are a mind tool that facilitate the drawing of conclusions (Chapter 4).

Innumeracy—ignorance of risk, miscommunication of risk, and clouded thinking—becomes a problem as soon as one is driven out of the promised land of certainty into the world in which Franklin's law reigns. Innumeracy, I emphasize, is not simply a problem within an individual mind; ignorance and miscommunication of specific risks can, for example, be produced and maintained by various groups within society to their own benefit.

Risk

The term "risk" has several senses. The one I intend has to do with uncertainty, but not necessarily regarding a dangerous event, such as a plane crash, because one can also be uncertain about a positive outcome, such as a successful landing. Another reason not to use the term to refer exclusively to negative outcomes is that there are situations in which a negative outcome from one perspective is a positive outcome from another. For instance, losing a month's salary in a gambling casino is a negative outcome for the gambler but a positive one for the casino.

In this book, I call an uncertainty a *risk* when it can be expressed as a number such as a probability or frequency on the basis of empirical data. The number need not be fixed—it may be updated in light of experience. In situations in which a lack of empirical evidence makes it impossible or undesirable to assign numbers to the possible alternative outcomes, I use the term "uncertainty" instead of "risk."[4] Uncertainty does not imply chaos; for instance, when a cure for cancer will be found is uncertain, but this does not have anything to do with chaos.

When does an uncertainty qualify as a risk? The answer depends on one's interpretation of probability, of which there are three major versions: *degree of belief, propensity,* and *frequency.* Degrees of belief are sometimes called *subjective probabilities.* Of the three interpretations of probability, the subjective interpretation is most liberal about expressing uncertainties as quantitative probabilities, that is, risks. Subjective probabilities can be assigned even to unique or novel events.

> *Degrees of Belief.* Consider the surgeon Christiaan Barnard's account of his first encounter with Louis Washkansky, who was then soon to become the first man to have a heart transplant. Washkansky was propped up in bed, reading a book. Barnard introduced himself and explained that he would exchange Washkansky's heart

for a healthy one and that "there's a chance you can get back to normal life again."[5] Washkansky did not ask how great the chance was, how long he would survive, or what the transplant operation involved. He just said he was ready to go ahead and turned back to his book, a Western. Barnard was deeply disturbed that Washkansky was more interested in pulp fiction than in this great moment in medical history and the risk it posed to him. But Washkansky's wife, Ann, did ask Barnard, "What chance do you give him?" Without hesitation or further explanation, he answered, "An 80 percent chance."[6] Eighteen days after the operation, Washkansky died. Barnard's "80 percent" reflected a degree of belief, or subjective probability. In the subjective view, uncertainties can always be transformed into risks, even in novel situations, as long as they satisfy the laws of probability—such as that the probabilities of an exhaustive and exclusive set of alternatives such as survival and death add up to 1. Thus, according to the subjective interpretation, Barnard's statement that Washkansky had an 80 percent chance of survival is meaningful provided that the surgeon also held that there was a 20 percent chance of his patient not surviving. In this interpretation, Barnard's "80 percent" would qualify as quantified uncertainty, that is, as risk.

Propensities. The possibility of translating uncertainties into risks is much more restricted in the propensity view. Propensities are properties of an object, such as the physical symmetry of a die. If a die is constructed to be perfectly symmetrical, then the probability of rolling a six is 1 in 6. The reference to a physical design, mechanism, or trait that determines the risk of an event is the essence of the propensity interpretation of probability. Note how propensity differs from the subjective interpretation: It is not sufficient that someone's subjective probabilities about the outcomes of a die roll are coherent, that is, that they satisfy the laws of probabilty. What matters is the die's design. If the design is not known, there are no probabilities. According to this view, Barnard's estimate of 80 percent would not qualify as a probability, or risk, because not enough is known about the heart operation for its propensities to be assessed.

Frequencies. For a frequentist, a probability must be based on a large number of observations and is defined as the relative frequency of an event in a specified *reference class,* such as the relative frequency of lung cancer in white American males who smoked cigarettes for at least 20 years. No reference class, no probability. Frequentists would not be interested in what someone *believes* about the outcome of rolling a die, nor would they need to study the *design* of the die to determine the probability of rolling a six. They would determine the probability empirically by rolling the die many times and computing the relative frequency with which the outcome was a six. Therefore, frequentists would declare Barnard's estimate of 80 percent meaningless (because there were no comparable transplants at the time he made the estimate), and hard-line frequentists would reject altogether the notion of assigning probabilities to a single event such as the survival of a specific man. Clearly, frequentists are cautious about moving from uncertainties to risks. For them, risks refer only to situations for which a large body of empirical data exists. The courts, for instance, tend to adhere to the frequentist position, admitting statements about risks as evidence only when they are based on empirical frequencies rather than opinion.

These different interpretations of probability can produce drastically different estimates of risk. A few years ago, I enjoyed a guided tour through Daimler-Benz Aerospace (DASA), which produces the Ariane, a rocket that carries satellites into orbit. Standing with my guide in front of a large poster that listed all 94 rockets launched so far (Ariane models 4 and 5), I asked him what the risk of an accident was. He replied that the security factor is around 99.6 percent. That was surprisingly high because on the poster I saw eight stars, which meant eight accidents. To be sure, several of the stars were next to early launchings, but launch numbers 63, 70, and 88 were also accidents. I asked my guide how eight accidents could translate into 99.6 percent certainty. He replied that DASA did not count the number of accidents, but rather computed the security factor from the design features of the individual parts of the rocket. He added that counting accidents would have included human errors, and pointed out that behind one of these stars, for instance, was a misunderstanding between one worker who had not installed a screw, and the worker on the next shift who had assumed that his predecessor had done so. The reported risk of an Ariane accident was, hence, based on a propensity, not a frequency, interpretation.

In this book, I will focus on risks that can be quantified on the basis of frequency data. This is not to say that frequencies are the whole story in estimating risks, but—when they are available—they provide a good starting point.

Ignorance of Risk

Who is informed about risks? The answer depends on one's culture and the event or hazard in question. For instance, the weather forecast might say that the chances of rain tomorrow are 30 percent, and we at least think we understand what that means. Although it seems natural to express the uncertainty of weather in terms of probabilities, this is a recent cultural phenomenon. Before 1965, the U.S. National Weather Service expressed its forecasts in all-or-none terms such as "it will not rain tomorrow," perhaps preceded by "it is unlikely that. . . ." In Germany, probabilities began to be reported in weather forecasts only around 1990; in France, weather forecasts are still largely probability-free. Some cultures have an insatiable appetite for numbers—batting averages, SAT scores, and market indices—while others are more reluctant to express uncertainties in numerical form. In general, democracies tend to have a greater desire for numbers and a greater motivation to make risks transparent than most other social systems.

PROMOTING PUBLIC IGNORANCE

However, democracies also host groups that have little interest in the public's knowing about certain risks. For instance, in the 1950s, the American tobacco industry began a massive campaign to convince the public that cigarette smoking was safe. This was around the time when the American scientific community began to reach a consensus that cigarettes are a major cause of illness, and the industry invested hundreds of millions of dollars in the creation of an illusion of certainty.[7] After the illusion crumbled following a report by the U.S. Surgeon General in 1964, the tobacco industry launched a second campaign of obfuscation to engender "doubt" about the extent of the actual risks involved. For decades, the scientific evidence concerning the hazards of smoking was rarely if ever discussed in the leading national magazines in which the tobacco industry advertised. Large segments of the public got the impression that the question of the effects of smoking on health was still open. As early as the mid-1950s, however, the American Cancer Society had evidence that people who smoked two packs of cigarettes a day were dying about seven years earlier, on average, than nonsmokers. Most experts today agree that tobacco is the cause of 80 to 90 percent of all cases of lung cancer. Tobacco kills upward of 400,000 Americans every year, primarily through lung cancer and heart disease; in Germany, the number is estimated to be 75,000. In China, the number of people who die from lung cancer will soon be close to 1 million a year. The case of cigarette smoking illustrates how public awareness of a health hazard can be diluted by a double defense line. First, the illusion of certainty is manufactured: Smoking is safe—period. When this illusion breaks down, uncertainty is acknowledged, but doubt is spread as to whether the actual risks are known or not.

BEYOND IGNORANCE:
IT'S OFTEN ONLY A SIMPLE CALCULATION

Not all ignorance is driven by trade lobbies or other parties that have an interest in keeping people ignorant of risks. There are also situations in which the facts are plainly in view and people have only to make a small mental effort to put them together.

What is the chance of one's dying in a motor vehicle accident over the course of a lifetime? It does not take much time to figure this out. In an average year, 40,000 to 45,000 people die on the roads in the United States.[8] Given that the country has about 280 million inhabitants, this means that about 1 in 7,000 of them is killed on the road each year. Assuming that this figure remains fairly stable over time, we can also figure out the chance of dying on the road during one's life. Given a life span of 75 years, the result is roughly 1 in 90. That is, 1 out of every 90 Americans will lose his or her life in a motor vehicle accident by the age of 75. Most of them die in passenger car accidents.

Are Americans in greater danger of being killed on the road than people in Germany or in Great Britain? In an average year, about 8,000 people die on the roads in Germany. Given its population of about 80 million, one can calculate that about 1 in 10,000 is killed in a motor vehicle accident. Over a life span of 75 years, this is equivalent to about 1 in every 130 people. Again, the majority of these people are killed while driving or riding in a passenger car.[9] Note that the higher fatality rate of Americans does not imply that they drive more dangerously than Germans; they just drive more, in part because of the lack of public transportation. In Great Britain (including Northern Ireland), the roads are safer than in the United States and Germany. There, again over a life span of 75 years, "only" about 1 in 220 people is killed in a motor vehicle accident.

American roads are, however, definitely not the most dangerous in the Western world. There are two European countries that stand out from the others, Portugal and Greece, where about 1 in 4,000 citizens is killed on the road every year. This means that over a life span of 75 years, about 1 out of every 50 people in Portugal and Greece is killed on the roads.[10]

All that is needed to make these estimates is the number of people who die of the cause in question each year and the population of the country. Both can be looked up easily for any country or state. These estimates are only rough because they do not take account of the possibility that driving behavior or safety technology might drastically change over time. I do not present the striking risks of driving to make every reader switch to public transportation. Many people have heard arguments of the sort "planes are safer than cars," yet these arguments do not change their behavior—because of habit, fear of flying, or love of driving. However, knowing the actual risk allows individuals to make up their own minds, to weigh the risk against the individual benefits of driving, and to arrive at an informed decision. For instance, the terrorist attack on September 11, 2001, cost the lives of some 3,000 people. The subsequent decision of millions to drive rather than fly may have cost the lives of many more.

PUBLIC NUMBERS

Public ignorance of risk has a historical basis. Unlike the stories, mythologies, and gossip that have been shaping our minds since the beginning of human culture, public statistics are a recent cultural achievement. During much of the eighteenth and nineteenth centuries, statistical information was a state secret known only by an elite and withheld from the public. The power of statistical information, such as population figures, has been recognized among political leaders for centuries. Napoleon's appetite for facts from his *bureau de statistique* was legendary.[11] And he always wanted the numbers immediately. At the Napoleonic court, the saying was, If you want something from Napoleon, give him statistics. Willingness to make economic and demographic figures public is a recent phenomenon. It was not until around 1830 that statistics, or at least some of them, became public. Since then, an "avalanche of printed numbers," to borrow the philosopher Ian Hacking's phrase, has turned modern life into a vast ocean of information conveyed by media such as television, newspapers, and the Internet. In this sense, one can say that, although uncertainties are old, risks are relatively new.

As already mentioned, the widening dissemination of statistical information to the public during the nineteenth and twentieth centuries has been linked to the rise of democracies in the Western world.[12] A democracy makes lots of information available to everyone, but its citizens often have very selective interests. It is more likely that a young American male knows baseball statistics than that his chance of dying on a motorcycle trip is about 15 times higher than his chance of dying on a car trip of the same distance.[13] Today, numbers are public, but the public is not generally numerate.

From Miscommunication to Risk Communication

The Prozac story in Chapter 1 illustrates the miscommunication of risk— the failure to communicate risk in an understandable way. Some forms of communication enhance understanding; others don't. Miscommunication of risk is often the rule rather than the exception and can be difficult to detect, as the ambiguous probabilities in the Prozac story illustrate. Statements about the probabilities of single events—such as "you have a 30 to 50 percent chance of developing a sexual problem"—are fertile ground for miscommunication. One mind tool that can overcome this problem is specifying a reference class, which occurs automatically when one uses statements about frequencies rather than single events.

There are three major forms of risk communication that invite miscommunication: the use of *single-event probabilities, relative risks,* and *conditional probabilities.* As it happens, these seem to be the most frequently used forms of risk communication today.

SINGLE-EVENT PROBABILITIES

To communicate risk in the form of a single-event probability means to make a statement of this type: "The probability that an event will happen is X percent." There are two reasons why such a statement can be confusing. First, as illustrated by the Prozac case, a probability of a single event, by definition, does not state what the reference class is. Second, if the event is unique, that is, there are no comparable events known, then the probability estimate itself is likely to be nothing but a wild guess that may suggest precision where, in fact, only uncertainty reigns. Let me give you some examples.

The statement "there is a 30 percent chance that it will rain tomorrow" is a probability statement about a singular event—it will either rain or not rain tomorrow. In contrast, the statement that it will rain on 10 days in May is a frequency statement. The latter statement can be true or false; a single-event probability by itself, however, can never be proven wrong (unless the probability is zero or one). Single-event probabilities can lead to miscommunication because people tend to fill in different reference classes. This happens even with such familiar statements as "there is a 30 percent chance that it will rain tomorrow." Some think this statement means that it will rain 30 percent of the time, others that it will rain in 30 percent of the area, and a third group believes it will rain on 30 percent of the days that are like tomorrow. These three interpretations are about equally frequent.[14] What weather forecasters actually have in mind is the last interpretation. However, people should not be blamed for different interpretations; the statement "there is a 30 percent chance that it will rain tomorrow" is ambiguous.

Dr. Barnard's 80 percent estimate illustrates specific problems with statements about unique events. Ann Washkansky may have gotten the impression that this high probability offered hope, but what it meant was ambiguous. Barnard did not say to what the number referred: the probability of Washkansky's surviving the operation, surviving the following day or year, or something else. Furthermore, the probability referred to the first heart transplant in history; there were no comparable cases on which Barnard could have based his estimate. Barnard's answer may have reassured, but did not inform, Washkansky's wife.

RELATIVE RISKS

What is the benefit of a cholesterol-lowering drug on the risk of coronary heart disease? In 1995, the results of the West of Scotland Coronary Prevention Study were presented in a press release: "People with high cholesterol can rapidly reduce . . . their risk of death by 22 per cent by taking a widely prescribed drug called pravastatin sodium. This is the conclusion of a landmark study presented today at the annual meeting of the American Heart Association."[15] The benefit of this cholesterol-reducing drug, just like that of most medical treatment, was reported by the press in the form of a *relative risk reduction*. What does "22 percent" mean? Studies indicate that a majority of people think that out of 1,000 people with high cholesterol, 220 of these people can be prevented from becoming heart attack victims.[16] This, however, is not true. Table 3-1 shows the actual result of the study: Out of 1,000 people who took pravastatin over a period of 5 years, 32 died, whereas of 1,000 people who did not take pravastatin but rather a placebo, 41 died. The following three presentations of the raw result—a total mortality reduction from 41 to 32 in every 1,000 people—are all correct, but they suggest different amounts of benefit and can evoke different emotional reactions in ordinary citizens.

TABLE 3-1 *Reduction in total mortality for people who take a cholesterol-reducing drug (pravastatin).* The people in the study had high-risk levels of cholesterol and participated in treatment for 5 years. (From Skolbekken, 1998.)

Treatment	Deaths (per 1,000 people with high cholesterol)
Pravastatin (cholesterol-reducing drug)	32
Placebo	41

Three Ways to Present the Benefit

Absolute risk reduction: The absolute risk reduction is the proportion of patents who die without treatment (placebo) minus those who die with treatment. Pravastatin reduces the number of people who die from 41 to 32 in 1,000. That is, the absolute risk reduction is 9 in 1,000, which is 0.9 percent.

Relative risk reduction: The relative risk reduction is the absolute risk reduction divided by the proportion of patients who die without treatment. For the present data, the relative risk reduction is 9 divided by 41, which is 22 percent. Thus, pravastatin reduces the risk of dying by 22 percent.

Number needed to treat: The number of people who must participate in the treatment to save one life is the number needed to treat (NNT). This number can be easily derived from the absolute risk reduction. The number of people who needed to be treated to save one life is 111, because 9 in 1,000 deaths (which is about 1 in 111) are prevented by the drug.

The relative risk reduction looks more impressive than the absolute risk reduction. Relative risks are larger numbers than absolute risks and therefore suggest higher benefits than really exist. Absolute risks are a mind tool that makes the actual benefits more understandable. Another mind tool serving as an alternative to relative risks is presenting benefits in terms of the number needed to treat to save one life. With this mind tool, one can see right away that out of 111 people who swallow the tablets for 5 years, 1 had the benefit, whereas the other 110 did not. The situation here is quite different from that of penicillin and other antibiotics whose positive effects when first introduced were dramatic.

CONDITIONAL PROBABILITIES

One can communicate the chances that a test will actually detect a disease in various ways (see Chapter 1). The most frequent way is in the form of a conditional probability: *If a woman has breast cancer, the probability that she will test positive on a screening mammogram is 90 percent.* Many mortals, physicians included, confuse that statement with this one: *If a woman tests positive on a screening mammogram, the probability that she has breast cancer is 90 percent.* That is, the conditional probability that an event A occurs given event B is confused with the conditional probability that an event B occurs given event A. This is not the only confusion. Others mistake the probability of A *given* B with the probability of A *and* B. One can reduce this confusion by replacing conditional probabilities with natural frequencies, as explained in the next chapter.

A RIGHT TO CLEAR INFORMATION

Despite the potential confusion created by single-event probabilities, relative risk reduction, and conditional probabilities, these forms of risk communication are standard. For instance, relative risks are the prevalent way in which the press and drug company advertising report the benefits of new treatments. There is a consensus today that the public has a right to information. But there is not yet a consensus that the public also has a right to get this information in a way that is clear and not misleading. I strongly urge medical, legal, and other associations to subscribe to an ethical policy that demands reporting risks in clear terms such as absolute risks and natural frequencies, rather than in ways that are more likely to confuse people. In this book, I introduce various mind tools for communicating risk in ways people can understand.

From Clouded Thinking to Insight

Ignorance of relevant risks and *miscommunication* of those risks are two aspects of innumeracy. A third aspect of innumeracy concerns the problem of *drawing incorrect inferences* from statistics. This third type of innumeracy occurs when inferences go wrong because they are clouded by certain risk representations. Such *clouded thinking* becomes possible only once the risks have been communicated. The mammography example in Chapter 1 illustrates a tool for achieving *mental clarity,* that is, a device for translating conditional probabilities—which impede not only risk communication but also correct inference from risks—into natural frequencies.

Why is it so difficult for even highly educated people to make inferences on the basis of probabilities? One reason might be that the theory of probability, which is concerned with drawing inferences from uncertain or incomplete information, is a relatively recent development in human history. Ian Hacking, who is fond of precise numbers, has dated this discovery to 1654, when the mathematicians Blaise Pascal and Pierre Fermat exchanged a now-famous series of letters about gambling. The fact that the notion of mathematical probability developed so late—later than most key philosophical concepts—has been called the "scandal of philosophy."[17] The difficulty that even great thinkers had in understanding risk before then is best illustrated by Girolamo Cardano, a sixteenth-century Italian physician and mathematician and the author of one of the first treatises on probability. Cardano, a notorious gambler, asserted that each face of a die will occur exactly once in any given six rolls. This assertion, however, flew in the face of his lifelong experience at the gambling tables. He resolved the conflict with an appeal to the intervention of luck (he was a great believer in his own). Cardano's intuition recalls that of the little girl who, as the story goes, was scheduled to receive an injection from her pediatrician. Upset that her father signed a consent form stating that he understood that 1 out of 10,000 children experience a serious allergic reaction, she insisted on speaking to the doctor. "I want to know," the little girl asked, "what number you're on."

The remainder of this book presents mind tools for overcoming innumeracy that are easy to learn, apply, and remember. I focus on three kinds of tools: Franklin's law for overcoming the illusion of certainty, devices for communicating risk intelligibly, and the use of natural frequencies for turning clouded thinking into insight. Overcoming innumeracy is like completing a three-step program to statistical literacy. The first step is to defeat the illusion of certainty. The second step is to learn about the actual risks of relevant events and actions. The third step is to communicate the risks in an understandable way and to draw inferences without falling prey to clouded thinking. The general point is this: Innumeracy does not simply reside in our minds but in the representations of risk that we choose.

Solving a problem simply means representing it so as to make the solution transparent.

Herbert A. Simon, *The Sciences of the Artificial*

4

INSIGHT

After leaving a restaurant in a charming town in Tuscany one night, I was looking for my yellow-green Renault in the parking lot. It wasn't there. Instead, I saw a blue Renault—the same model but the wrong color. I can still feel my fingers hesitating to put the key into the lock, but the door of the blue car opened. I drove the car home. When I looked out the window the next morning, I saw a yellow-green Renault standing in bright sunlight outside. What had happened? My color constancy system had failed in the parking lot's artificial light but was functioning correctly under the next day's sun. Color constancy, an impressive adaptation of the human perceptual system, allows us to see an object as having the same color under diverse conditions of natural illumination—for instance, in the bluish light of day as well as in the reddish light of sunset. Under conditions of artificial illumination, such as that produced by sodium or mercury vapor lamps, color constancy can break down.

Human color vision is adapted to the spectral properties of natural sunlight. More generally, our perceptual systems have been shaped by the environment in which our ancestors evolved, often referred to as the "environment of evolutionary adaptedness."[1] These adaptations can be exquisite, though not foolproof. The human visual system is far better than any camera at registering constant colors with changing illumination. Similarly, human morphology and physiology and the human nervous and immune systems all reflect ingenious adaptations. The tubular structure of bones, for instance, maximizes strength and flexibility while minimizing weight; pound for pound, bones are stronger than solid steel bars. The best manmade heart valves cannot match the way natural heart valves open and close. Like color constancy, however, these adaptations may break down—in the case of bones, even in the literal sense—when stable properties of the environment to which they are adapted change.[2]

We can understand innumeracy by using the analogy of failures of color constancy. Just as certain types of illumination can enhance or interfere with color constancy, certain numerical representations can help or hinder sound statistical thinking. In my view, the problem of innumeracy is not essentially "inside" our minds as some have argued, allegedly because the innate architecture of our minds has not evolved to deal with uncertainties. Instead, I suggest that innumeracy can be traced to external representations of uncertainties that do not match our mind's design—just as the breakdown of color constancy can be traced to artificial illumination. This argument applies to the two kinds of innumeracy that involve numbers: miscommunication of risks and clouded thinking. The treatment for these ills is to restore the external representation of uncertainties to a form that the human mind is adapted to.

A Physician's Thinking

Dr. Konrad Standing[3] is chief of a department in a university teaching hospital, a prominent figure in research and teaching with more than three decades of professional experience. A few years ago, I asked him whether the physicians at his institution would participate in a study of diagnostic intuitions being conducted by my colleagues and me at the Max Planck Institute for Human Development. Seeming quite interested in the subject, he agreed to encourage his colleagues to participate. To set an example, he volunteered himself. The first diagnostic task he worked on concerned the routine breast cancer screening of women, as described briefly in Chapter 1:

To facilitate early detection of breast cancer, starting at a particular age, women are encouraged to participate at regular intervals in routine screening, even if they have no obvious symptoms. Imagine that you conduct such breast cancer screening using mammography in a particular region of the country. The following information is available about asymptomatic women aged 40 to 50 in such a region who participate in mammography screening:

The probability that one of these women has breast cancer is 0.8 percent. If a woman has breast cancer, the probability is 90 percent that she will have a positive mammogram. If a woman does not have breast cancer, the probability is 7 percent that she will still have a positive mammogram. Imagine a woman who has a positive mammogram. What is the probability that she actually has breast cancer?[4]

Department chiefs are not used to having their reasoning tested. Dr. Standing was visibly nervous while trying to figure out what he would tell the woman. After mulling the numbers over, he finally estimated the woman's probability of having breast cancer, given that she has a positive mammogram, to be 90 percent. Nervously, he added, "Oh, what nonsense. I can't do this. You should test my daughter; she is studying medicine." He knew that his estimate was wrong, but he did not know how to reason better. Despite the fact that he had spent 10 minutes wringing his mind for an answer, he could not figure out how to draw a sound inference from the probabilities.

If your mind, like Dr. Standing's, is clouded by this problem, don't despair. This feeling is at the crux of the point I would like to demonstrate. Innumeracy? Yes, arising from clouded thinking. Treatment? The same as for failures of color constancy.

When color constancy fails under sodium vapor lamps, the solution lies outside of the mind, not inside. One needs to restore the type of input that the brain has encountered during most of its evolution and to which the color constancy mechanism is therefore adapted: sunlight. In this analogy, probabilities are like sodium vapor lamps. What then corresponds to natural light in the mammography problem? I propose that the answer is *natural frequencies,* that is, simple counts of events.

Let us try to turn Dr. Standing's innumeracy into insight by communicating in natural frequencies rather than probabilities:

> *Eight out of every 1,000 women have breast cancer. Of these 8 women with breast cancer, 7 will have a positive mammogram. Of the remaining 992 women who don't have breast cancer, some 70 will still have a positive mammogram. Imagine a sample of women who have positive mammograms in screening. How many of these women actually have breast cancer?*

The information is the same as before (with rounding) and it leads to the same answer. But now it is much easier to see what that answer is. Only 7 of the 77 women who test positive (70 + 7) actually have breast cancer, which is 1 in 11, or 9 percent—much lower than the estimate of 90 percent that Dr. Standing had given. When he received the information in natural frequencies, his innumeracy turned into insight. With frequencies, he "saw" the answer, remarking with relief, "But that's so easy" and even "That was fun." His daughter's help was no longer needed.

How to Turn Physicians' Innumeracy into Insight

Do natural frequencies help to clear physicians' minds in general, as they did for Dr. Standing? Ulrich Hoffrage and I tested 48 physicians with an average of 14 years of professional experience from two major German cities. About two-thirds of these physicians worked in private, public, or university hospitals, and the rest in private practice. They included radiologists, gynecologists, internists, and dermatologists, among others. Their professional status ranged from being fresh out of medical school to heading their respective medical departments. As with Dr. Standing, we asked each physician to estimate the chances of breast cancer in a woman aged 40 to 50 given a positive mammogram in a routine screening. Half of the physicians received the relevant information in probabilities; the other half received the same information in natural frequencies.[5]

With probabilities, there was alarmingly little consensus, as the left side of Figure 4-1 shows. The estimates ranged between 1 percent and 90 percent. Like Dr. Standing, one-third of the physicians (8 out of 24) concluded that the probability of breast cancer given a positive mammogram is 90 percent. Another third of physicians estimated the chances to be between 50 percent and 80 percent. Another 8 thought the chance was 10 percent or less, and half of these estimated the probability as 1 percent, which corresponds to the base rate. The median estimate was 70 percent. If you were a patient, you would be justly alarmed by this diversity of medical opinion. Only 2 of the physicians reasoned correctly, giving estimates of about 8 percent. Another 2 estimated the chances to be near this value, but for

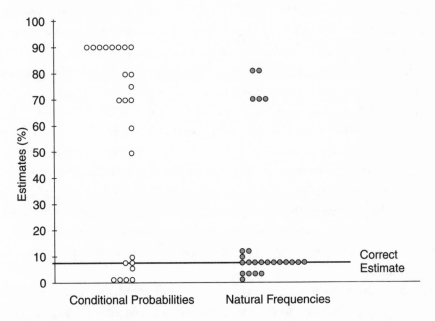

FIGURE 4-1. *Estimated chances of breast cancer given a positive screening mammogram.* Of 48 physicians, half received the relevant information in conditional probabilities, the other in natural frequencies. Each point represents one physician. The ordinate shows their estimates of the probability or frequency of breast cancer after a positive test. With probabilities, the physicians were highly inconsistent; with natural frequencies, this inconsistency largely disappeared—except for five "hopeless" cases—and the physicians' estimates clustered around the correct one. (From Gigerenzer, 1996a; Hoffrage and Gigerenzer, 1998.)

the wrong reasons. For instance, 1 physician confused the false positive rate with the probability that the patient has breast cancer given that she has a positive mammogram.

When the information was presented in probabilities, the majority of physicians in our study grossly overestimated the risk of breast cancer given a positive mammogram, just as Dr. Standing did. They did so despite having spent considerable time thinking about the problem. It was evident that they didn't take the test lightly—many of them were nervous and tense about being tested, a situation that they rarely, if ever, encounter.

How did physicians manage with natural frequencies? The right side of Figure 4-1 shows that the disquieting variability in responses largely disappeared. The majority of the physicians in this group responded with the correct answer, or close to it. Only five of the physicians who received the information in natural frequencies concluded that the chance of breast cancer given a positive mammogram was above 50 percent. Simply stating the information in natural frequencies turned much, though not all, of the physicians' innumeracy into insight.

Is clouded thinking specific to the German physicians in our study? It seems not. David Eddy, former consultant to the Clinton administration on health care reform, asked a number of American physicians to estimate the probability that a woman has breast cancer given that she has a positive mammogram, providing essentially the same information as we did.[6] Eddy gave all the physicians the relevant information in probabilities, not in natural frequencies. Ninety-five out of 100 of them estimated the probability of breast cancer to be about 75 percent, about 10 times more than the realistic estimate.

I am grateful to the physicians who volunteered to participate in our study. They enabled us to demonstrate, for the first time, that frequency representations can help experienced physicians make better diagnostic inferences. The implication of this finding is not to blame physicians' (or patients') inability to reason about probabilities. Rather, the lesson is to represent risks in medical textbooks and in physician-patient interactions in a way that comes naturally to the human mind. Natural frequencies are a simple, inexpensive, and effective method of improving diagnostic insight.

Insight from Outside

Why does representing information in terms of natural frequencies rather than probabilities or percentages foster insight? For two reasons. First, computational simplicity: The representation does part of the computation. And second, evolutionary and developmental primacy: Our minds are adapted to natural frequencies.

THE REPRESENTATION DOES PART OF THE COMPUTATION

Figure 4-2 illustrates the difference between natural frequencies and probabilities. On the left side is a tree with natural frequencies, which represents

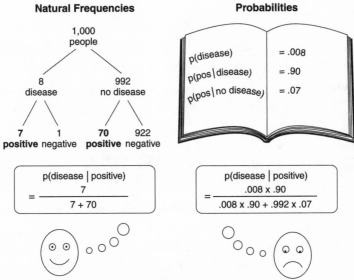

FIGURE 4-2. *How natural frequencies facilitate Bayesian computations.* The happy person received the relevant information in natural frequencies and has an easy time estimating the chances of disease given a positive test (or symptom). The reason is that she only has to pay attention to two numbers, the number of patients with a positive test and the disease ($a = 7$) and the number of patients with a positive test and no disease ($b = 70$). The person with the unhappy face has received the same information in probabilities and has a hard time making this estimation. The structure of his equation is exactly the same as the one the happy person used—$a/(a + b)$—but the natural frequencies a and b have been transformed into conditional probabilities, making the formula for probabilities much more complex.

how a person would encounter statistical information through direct experience. On the right side is the same information in probabilities, the format in which most information is represented for medical students in textbooks. The numbers are the same as those in the breast cancer problem that Dr. Standing struggled to solve. The thought bubbles show the calculations needed to answer the question we asked in each case.

Both equations are versions of Bayes's rule,[7] which is named after the English dissenting minister who is thought to have discovered it, Reverend Thomas Bayes (1702(?)–1761).[8] You can see that calculating the probability of a disease given a positive test is easier when the information is in natural frequencies:

$$p(\text{disease}|\text{positive}) = \frac{a}{a+b}$$

Bayes's rule for natural frequencies

In Figure 4-2, *a* is the number of people who test positive *and* have the disease (7) and *b* is the number of people who test positive and do *not* have the disease (70).[9] With probabilities, in contrast, the calculation is more demanding:

$$p(\text{disease}|\text{positive}) = \frac{p(\text{disease})p(\text{positive}|\text{disease})}{p(\text{disease})p(\text{positive}|\text{disease}) + p(\text{no disease})p(\text{positive}|\text{no disease})}$$

Bayes's rule for conditional probabilities

This rule is the same as the simpler one above; both show the proportion of correct positives (numerator) among all positives (denominator). The difference in this version of the rule is that each of the natural frequencies has been replaced by the product of two probabilities. Table 4-1 explains these probabilities.

In general, a test has four possible outcomes. When a person has a disease, the test can be either positive (a *true positive*) or negative (a *false negative*). The probability $p(\text{positive}|\text{disease})$ is the *sensitivity* of the test. The sensitivity of mammography is the proportion of women who test positive

TABLE 4-1. *Test outcomes.* A test can have four possible outcomes: (*a*) a positive result given disease (or some other unknown condition), (*b*) a positive result given no disease, (*c*) a negative result given disease, and (*d*) a negative result given no disease. The rates with which these four results occur are called (*a*) sensitivity (or true positive rate), (*b*) false positive rate, (*c*) false negative rate, and (*d*) specificity (true negative rate). The two shaded areas indicate the two possible errors. The frequency of true positives and false positives are *a* and *b*, respectively, in Bayes's rule.

Test Result	Disease	
	Yes	No
Positive	(*a*) sensitivity	(*b*) false positive rate
Negative	(*c*) false negative rate	(*d*) specificity

among those who have breast cancer. It usually ranges between 80 percent and 95 percent, with lower values among younger women. The rates of these two outcomes—the sensitivity and the false negative rate—add up to 1.

When a person does not have the disease, the test can be either positive (a *false positive*) or negative (a *true negative*). The rates of these two outcomes, the false positive rate and the specificity, also add up to 1. The probability $p(\text{positive}|\text{no disease})$ is the false positive rate of a test. The false positive rate of mammography is the proportion of women who test positive among those who do not have breast cancer. It ranges between about 5 and 10 percent, with the higher values among younger women.

The two of the four possible outcomes that are errors are shaded in Table 4-1. The rates of the two errors are dependent on one another: Decreasing the false positive rate of a test increases the false negative rate, and vice versa. The four probabilities in Table 4-1 are called *conditional probabilities,* because they express the probability of some event (for example, a positive test) conditional on the occurrence of some other event (for example, disease)—that is, *given* that this other event has occurred. The non-conditional probability p(disease) is the *base rate* of having the disease. That is, the base rate is the proportion of individuals with a certain disease in a particular population at a specific point in time. In contrast to the base rates, conditional probabilities are notoriously confusing.

We can now understand exactly why this is the case. When natural frequencies are transformed into conditional probabilities, the base rate information is taken out (this is called *normalization*). The benefit of this normalization is that the resulting values fall within the uniform range of 0 and 1. The cost, however, is that when drawing inferences from probabilities (as opposed to natural frequencies), one has to put the base rates back in by multiplying the conditional probabilities by their respective base rates.[10]

To summarize: Natural frequencies facilitate inferences made on the basis of numerical information. The representation does part of the reasoning, taking care of the multiplication the mind would have to perform if given probabilities. In this sense, *in*sight can come from *out*side the mind.

MINDS ARE ADAPTED TO NATURAL FREQUENCIES

Natural frequencies result from natural sampling, the process by which humans and animals have encountered information about risks during most of their evolution. In contrast, probabilities, percentages, and other normalized forms of representing risk are comparatively recent. Animals can monitor the frequencies of important events fairly accurately. Rats, for instance, are able to "count" up to about 16, as evidenced by their ability to learn to press a lever a fixed number of times in order to get a dose of food.[11] Humans can also monitor frequencies fairly accurately, although not as accurately as David Hume believed when he claimed that humans can detect the difference between an event that occurs 10,000 times and one that occurs 10,001 times intuitively, that is, without externally recording the number of occurrences. The human mind records the frequencies of events, like the spatial and temporal locations of objects, with little effort, awareness, or interference from other processes.[12] Human babies are reportedly able to discriminate between groups of one, two, or three objects (such as black dots and Mickey Mouse toys) just a few days after birth.[13] Studies of counting in children also indicate that intuitions about numbers naturally focus on discrete cases rather than fractions (such as conditional probabilities). For instance, if 3- or 4-year-old children are shown five forks, one of which is broken into two parts, and are asked how many forks they can see, most say six.[14]

Like children, mathematicians began by thinking in terms of frequencies and only later turned to fractions, percentages, and probabilities. (For Pythagoras and his followers, numbers meant positive integers, not fractions or negative numbers. According to legend, the mathematician Hippasus of Metapontum was thrown off a ship for proving the existence of irrational numbers and thereby shattering the Pythagorean view that the world is ruled by integers.)[15] Probabilities and percentages are historically recent representations of uncertainty. The mathematical theory of probability emerged only in the mid–seventeenth century. Percentages became common notations during the ninteenth century after the metric system was instituted in Paris following the French Revolution, but were mainly used to express interest rates and taxes rather than uncertainty.[16] Only in the second half of the twentieth century did probabilities and percentages become entrenched in everyday language as expressions of uncertainty, such as in weather reports and baseball statistics. Over the course of most of its evolution, the human mind did not learn about risks from probabilities or percentages.

Representation Matters

Good numeric representation is a key to effective thinking that is not limited to understanding risks. Natural languages show the traces of various attempts at finding a proper representation of numbers. For instance, English does not denote numbers consistently in the base-10 system. Instead, it has special names for the numbers 1 to 12 that are vestiges of an earlier base-12 system. This base-12 system was also used for units of money and length—for example, there were 12 pennies to a shilling and there are 12 inches to a foot. Children in English-speaking countries also have to learn special words for the numbers between 13 and 20, such as "thirteen" and "fourteen," which hearken back to an earlier base-20 system. French children have to deal with other remnants of the base-20 system, such as *quatre-vingt-dix* ("four twenty ten"), which means 90. German-speaking children, like their English-speaking counterparts, have to learn a similar hodge-podge of words for the numbers up to 20—and face an additional challenge. In German, the written sequence of two-digit numbers is reversed when spoken, for example, the word *vierundzwanzig* ("four and twenty") means 24. Chinese, in contrast, denotes numbers consistently in the base-10 system. For instance, the Chinese words for two-digit numbers can be generated by a simple rule: 11 is expressed as "ten one," 12 as "ten two," and so on; 20 is expressed as "two ten," 21 as "two ten one," and so on.

Different linguistic representations of numbers can speed up or slow down learning for speakers of different languages. The psychologist Kevin Miller and his colleagues asked matched groups of American and Chinese children to recite the counting sequence from the number 1.[17] At age four, the American children could count up to 15, on average, whereas the Chinese children could count up to 40. One reason for the relatively poor performance of American children seems to be that two-digit numbers are less transparently represented in English than in Chinese.

The key role of representation in thinking is often downplayed because of an ideal of rationality that dictates that whenever two statements are mathematically or logically the same, representing them in different forms should not matter. Evidence that it does matter is regarded as a sign of human irrationality. This view ignores the fact that finding a good representation is an indispensable part of problem solving and that playing with different representations is a tool of creative thinking. The physicist Richard Feynman observed that different representations of the same physical law, even if they are mathematically equivalent, can evoke different mental images—and trigger new discoveries as a result.[18]

External representations are not just passive inputs to an active mind. They can do part of the reasoning or calculation, making relevant properties of the same information more accessible. In an arabic-numeral representation, for instance, one can quickly see whether a number is a power of 10, but not whether it is a power of 2. The converse holds for binary numbers. In addition, arabic numerals are better suited to multiplication and division than roman numerals—a difference that may help to explain the superior development of mathematics in early arabic cultures relative to Roman and early medieval European culture. Imagine trying to multiply the roman numbers XIX and XXXIV. In the arabic system, one simply multiplies each of the two digits in the first number (19) by each digit in the second number (34) and writes the results such that consecutive digits represent 1s, 10s, 100s, and so on. This is impossible using the roman numerals—although they also constitute a base-10 system—because each digit does not denote a power of 10. As a result, the roman-numeral system does not facilitate multiplication and division.

In this chapter, I have described a tool that can help to defog minds: changing risk representations from probabilities into natural frequencies. Probabilities—or more precisely, conditional probabilities—tend to impede human inference, whereas natural frequencies demand less computation and have a form in which the human mind has experienced events over most of its evolution. As the case of Dr. Standing illustrates, natural frequencies help not only laypeople, but experts as well. Restoring this "original" representation can turn innumeracy into insight.

PART II

UNDERSTANDING UNCERTAINTIES
IN THE REAL WORLD

I hope very much that pressure is not put on women to attend [breast cancer screening]. The decision must be theirs, and a truthful account of the facts must be available to the public and the individual patient. It will not be what they want to hear.

M. Maureen Roberts,
Clinical director, Edinburgh Breast Cancer Screening Project

5

BREAST CANCER SCREENING

Shortly before dying of breast cancer, the clinical director of an initiative called the Edinburgh Breast Cancer Screening Project, Maureen Roberts, wrote of mammography screening, "We can no longer ignore the possibility that screening may not reduce mortality in women of any age, however disappointing that may be." She went on to remark on what she called the "brainwashing" of physicians and the public: "There is also an air of evangelism [in the national screening programs], few people questioning what is actually being done."[1] Dr. Roberts was referring to mammography screening, not to the use of mammography per se. Mammography screening is mass screening, primarily of healthy women. In this situation, statistical thinking becomes of the utmost relevance, as we will see. Mammography is also used in other situations, for instance for patients with known symptoms of breast cancer, such as a tumor mass detected in a clinical exam. The benefits and hazards of mammography screening cannot be directly generalized to these other situations.

Mass screening for breast cancer was first introduced in Germany in the

1930s. The reason was not so much that X rays were discovered in Germany, but rather that Germans were among the first to recognize the cancer hazards of cigarettes, asbestos, tar, and radium, prompted by the unions and political parties that represented the interests of the working class.[2] Physicians were exhorted to recognize the value of early detection, and those who did not perform mammography screening were accused of being complicit in the deaths of thousands of women each year. Radio and newspaper announcements urged women over the age of 30 to participate in annual or biennial screening. Hundreds of thousands were screened using mammograms, and monthly breast self-examination was declared to be the moral duty of every woman. World War II, however, put a damper on German physicians' hopes of combating cancer. Comparable American campaigns did not get started until 30 years later.

More than 10 years after Dr. Roberts expressed her professional opinion about screening, women in North America are still getting contradictory recommendations.[3] The disagreements concern each of the three screening tests: mammography, clinical breast examination, and breast self-examination. The main disagreement about mammography has to do with the age at which it has a benefit for women. The American Cancer Society and the National Cancer Institute recommend that women have mammograms and clinical breast exams annually or every one to two years, respectively, beginning at age 40. The U.S. Preventive Services Task Force and the Canadian Task Force on the Period Health Exam, in contrast, recommend mammography screening every one to two years beginning at age 50. The main disagreement about the clinical and self-exams concerns whether they should be recommended at all. The U.S. Preventive Services Task Force asserts that mammography screening may as well be performed without the clinical exam because, it argues, there is no evidence that the clinical exam increases the benefit of mammography, whereas the other three organizations recommend the clinical breast exam as well. Finally, the American Cancer Society also recommends that women perform self-exams every month starting at age 20. None of the other organizations recommends self-exams at any age.

Given these different recommendations, which ones, if any, should women follow? To answer this question, they need to be well informed about the costs and benefits of screening. And this decision can vary by individual, because the same benefits and hazards have different values for different women.

Knowledge concerning screening is less universal than one might believe. For example, in surveys, more than one-third of a sample of African American women in south Florida and one-third of a sample of Hispanic women in Washington State reported never having heard of mammography. Others do not understand the nature of screening. In a survey of Australian women and men, 4 out of 5 did not know that screening tests are for asymptomatic people.[4] Five of the most common confusions about screening are listed below.

Clarifying Five Common Misunderstandings

- *Are screening tests meant for patients with known symptoms?* No, screening is for asymptomatic people. Its purpose is early detection.

- *Does screening reduce the incidence of breast cancer?* No, it does not. Early detection does not amount to prevention.

- *Does early detection imply mortality reduction?* Not in all cancers. Early detection can, but need not, lead to mortality reduction. For instance, if there is no effective therapy, early detection has no effect on mortality. Here, early detection does not increase life expectancy, but only the time the patient consciously has to live with the cancer.

- *Do all breast cancers progress?* No. Mammography can detect a form of breast cancer that is called "ductal carcinoma in situ." Most cancers found in the screening of younger women are ductal carcinomas in situ. Although their clinical course is not well understood, half or more of the lesions do not seem to progress.

- *Is early detection always a benefit?* No. For instance, when a cancer does not progress or progresses so slowly that it has no effect during a woman's life, early detection does not help this woman. Rather, she will suffer harm without benefits: Most likely she will undergo invasive treatment, such as mastectomy or lumpectomy with radiation, and her quality of life will substantially decrease.

These basic misunderstandings aside, what are the costs and benefits, both real and perceived, of mammography screening?

Benefits

The goal of mammography screening is to reduce mortality from breast cancer by early detection; screening cannot prevent cancer. How can one determine whether and when screening actually serves that goal? The best evidence comes from *randomized trials,* in which a large number of women are randomly assigned to either a screening group or a control group. Women in the screening group participate in screening at regular intervals. They are called back if they have a positive mammogram, and additional mammograms or biopsies are performed. And if they are diagnosed with breast cancer, they receive treatments such as a mastectomy (total removal of the breast), a lumpectomy (removal of only the affected tissue), and radiation therapy. Women in the control group do not participate in screening. After a fixed period, say 10 years, the two groups are compared to determine how many lives screening actually saved. Because the participants are randomly assigned, any difference in mortality that is observed can be attributed to the screening rather than to initial differences between the groups on dimensions such as age, social status, or health.

Ten large randomized trials have been conducted to determine whether undergoing mammography screening decreases women's chances of dying from breast cancer. These trials involved a total of half a million women in Canada, Scotland, Sweden, and the United States.[5] What were the results?

HOW TO COMMUNICATE BENEFITS?

The results will look different, and may even be misunderstood, depending on how they are presented. Consider first the overall benefit of mammography screening across all age groups (age 40 and older) covered by four Swedish randomized trials.[6] Table 5-1 shows the results of the screening.

Four Ways to Present the Risk

Relative risk reduction. One way to present the risk is to say that mammography screening reduces the risk of dying from breast cancer by 25 percent, that is, the relative risk reduction is 25 percent. Health organizations favor reporting benefits in terms of relative risk, thereby opening the door to miscommunication. What does a relative risk reduction of 25 percent mean? Many believe incorrectly that this means that out of 100 women who participate in screening, the lives of 25 will be saved. Yet to understand what relative risks mean, let us translate them into the mind's hard currency, that is, concrete cases. Consider 1,000 women who did not participate in screening and another 1,000 who did (Table 5-1). Within 10 years, 4 women in the first group and 3 in the second group died of breast cancer.[7] The decrease from 4 to 3 is a relative risk reduction of 25 percent.

Absolute risk reduction. The absolute risk reduction is 4 minus 3, that is, 1 out of 1,000 (which corresponds to 0.1 percent). In other words, if 1,000 women participate in screening for 10 years, one of them will probably be saved from dying of breast cancer.

Number needed to treat. There is a third way to communicate the benefit of screening: the *number needed to treat* in order to save one life. The smaller this number is, the better the treatment. In the present case, this number is 1,000, because this many women must participate in order to save one life.

Increase in life expectancy. Finally, one can express the benefit as an *increase in life expectancy.* Women who participate in screening from the age of 50 to 69 increase their life expectancy by an average of 12 days.[8]

TABLE 5-1: *Reduction in breast cancer mortality for women (age 40 and above) over a period of 10 years.* The raw result is from four Swedish randomized trials with some 280,000 women (numbers rounded). (Data from Nyström et al., 1996, in Mühlhauser and Höldke, 1999.)

Treatment	Deaths (per 1,000 women)
No mammography screening	4
Mammography screening	3

All four presentations of the raw data are correct, but they suggest different amounts of benefits and evoke different degrees of willingness to participate in screening and different emotional reactions in women. When risk reduction is expressed in relative terms, misunderstanding is likely. Relative risks are bigger numbers than absolute risks—compare, for instance, a 25 percent relative risk reduction and the absolute risk reduction of 1 life in 1,000. Counting on their clients' innumeracy, organizations that want to impress upon clients the benefits of a treatment generally report them in terms of relative risk reduction. Relative risks do not carry information about the absolute benefits of a treatment: For example, a 25 percent reduction means many lives saved if the disease is frequent, but only a few if the disease is rare. Transparent risk communication can be achieved by means of absolute risks, number needed to treat, and increases in life expectancy. But health organizations rarely use these, communicating with the public in terms of relative risk reduction instead. In Chapter 12, we will see that there are often institutional reasons for presenting benefits in terms of relative risks; for instance, given the innumeracy of health authorities who evaluate proposals for medical research funding, applicants often feel compelled to report relative risk reductions because they sound more impressive.

Transparency can also be achieved by expressing the benefit of a treatment in terms of a more familiar situation. For instance, participating in annual mammography screening that affords a 25 percent risk reduction has roughly the same effect on life expectancy as reducing the distance one drives each year by 300 miles.[9]

DOES SCREENING BENEFIT WOMEN IN THEIR 40S?

None of the 10 randomized trials suggested that mammography screening of women aged 40 to 49 reduces breast cancer mortality in the nine years following the beginning of screening. Nine out of the 10 trials also found no evidence of a mortality reduction after 10 to 14 years, although one— conducted in Gothenburg, Sweden—did. Why this trial found a reduction while the others did not remains unclear. This study was not specifically designed for women in their 40s, whereas the Canadian National Breast Cancer Study—the only trial designed for this age group, and including twice as many women as the Gothenburg trial—did not find a mortality reduction after 10.5 years. When the results of all 10 randomized trials were pooled, no reduction in breast cancer mortality was detectable even after 10 to 14 years.[10] Thus, there is yet no evidence that mammography screening reduces mortality from breast cancer for women in their 40s.

Why would screening women under 50 fail to reduce mortality from breast cancer? There are some possible reasons, but no conclusive answers. For instance, it has been posited that breast density is generally higher in younger women, making screening less likely to detect cancers at a curable stage. An alternative explanation is that, in younger women, a greater proportion of invasive breast cancers are aggressive and therefore grow more rapidly, which results in more cancers not detected between regular screening examinations. In addition, women under 50 are less likely to have breast cancer, which means that fewer women in this group can benefit from screening in the first place.[11]

DOES SCREENING BENEFIT WOMEN 50 AND OVER?

Eight of the 10 randomized clinical trials included women who began screening at or over age 50. Three of these studies found a significant reduction in breast cancer mortality; four found reductions that were too small to be distinguishable from zero; and one found no reduction at all. The reductions in mortality rates, which were computed seven to nine years after the first screening, appeared as early as four years after the first screening. When the results of all the trials are pooled, one finds a relative risk reduction of about 27 percent. But how many lives are saved, that is, how large is the absolute risk reduction? Consider women who start screening at age 50 and have a mammogram every other year for the next 20 years. For every 270 of these women, the life of one is saved. Thus, the absolute risk reduction is 1 in 270, or about 4 in 1,000. Still, the benefits are greater in this age group than across all age groups (Table 5-1). From age 50 on, mammography screening seems to reduce mortality from breast cancer.[12]

ARE CLINICAL BREAST EXAMS
AND SELF-EXAMS WORTHWHILE?

Common sense suggests that using all three methods of screening for breast cancer would be better than using just one. However, this does not seem to be the case. Among women aged 50 and older, the clinical breast exam does not contribute to decreasing breast cancer mortality compared to mammography alone. On the other hand, mammography contributes little to the benefit of the clinical breast exam if the exam is performed by a skilled practitioner.[13] Similarly, several studies with women between the ages of 35 and 65 found that performing regular self-exams had *no* effect on breast cancer mortality, despite increasing the number of breast cancers detected. Yet the costs of clinical exams and self-exams can be high, because women who consult an oncologist after finding a suspicious change may go through months or even years of physical and psychological strain without any benefit, as described below. Self-exams, which are recommended on a monthly rather than a yearly basis, are particularly likely to generate a steady stream of suspicion and anxiety, and many women obtain mammograms to gain peace of mind.[14]

SUMMARY OF BENEFITS

On the benefit side, the situation does not seem as bleak as Maureen Roberts described it. There is evidence that mammography screening reduces mortality from breast cancer in women aged 50 and older. As a consequence, early detection in this age group might also reduce the chance of invasive therapy and improve quality of life. But the situation for women in their 40s is unclear. There is currently no evidence that there is a benefit before 10 years after the beginning of screening. There is also no evidence that the clinical breast exam or self-exam has any additional benefit when performed in addition to mammography screening. These findings have led to the reversal of earlier recommendations made by health organizations with respect to the best way to screen for breast cancer. For instance, whereas 10 years ago health organizations still recommended that women between 35 and 39 have a baseline mammogram, today no responsible organization recommends baseline mammograms or screening of women in their 30s.

Finally, does annual screening for breast cancer result in a higher risk reduction than biennial screening? No. Whether women underwent mammography annually or biennially made no difference in the randomized trials[15]—apparently because the length of time that must pass before many tumors become detectable by mammography is around 3.5 years, which leaves enough time for biennial screening to detect cancer.

In a recent consensus statement, the National Institutes of Health put the decision explicitly into the hands of patients: "The available data do not warrant a blanket recommendation for mammography for all women in their forties. Each woman should decide for herself whether to undergo mammography."[16] This statement has upset many people who expect firm guidelines. In the past, only a few women seem to have decided to undergo screening of their own accord; most followed their physician's recommendation. When the physician recommended a screening mammogram, some 90 percent complied; when the physician did not, only 10 percent of women had screening mammograms.[17]

How can a woman make up her own mind about screening? To make a decision independently and wisely, she needs to know mammography's potential costs in addition to its benefits.

Costs

The costs of mammography screening are not as well documented as their benefits. These costs include physical and psychological harm and financial expense. Three groups of women pay the costs incurred by breast cancer screening.[18]

FALSE POSITIVES

The first group is made up of women who do *not* have breast cancer, but have a *positive* mammogram (that is, a false positive). Women in this group are called back for further investigation. Almost all of them are then subjected to another mammogram, an ultrasound, or a biopsy, and an unlucky few then undergo a lumpectomy or mastectomy. For many women, mammograms are painful and upsetting. For some, mammography screening is psychologically traumatic and results in an extended period of anxiety or depression and a loss of concentration. Biopsies, too, can have psychological costs, as well as physical costs such as wound infections, hematomas, scarring, and the loss of breast tissue. While some women might be grateful at first to find out that the result was a *false* positive, the emotional shock of a positive mammogram and its follow-up sometimes persists even after a negative biopsy. Three months after receiving false positive mammograms, 1 out of 2 women reported feeling considerable anxiety about mammograms and breast cancer. Moreover, 1 in 4 women reported that this anxiety affected their daily mood and functioning.[19]

How large is this group of women? In other words, how common are false positive mammograms?

> *First mammogram.* In an investigation of 26,000 women who underwent mammography screening for the first time, only 1 out of 10 who tested positive was found to have breast cancer at some point during the 13-month period thereafter.[20] In other words, 9 out of 10 positive results later proved to be false positives. Figure 4-2 (page 45) illustrates this result: For every 1,000 women who have their first mammogram between 40 and 50 years of age, 70 of them can expect to receive false positives compared with only 7 true positives. Among younger women, the proportion of false positives among all positives is even higher.[21]

Repeated mammograms. What about women who undergo mammography regularly, that is, annually or biennially? After undergoing a series of 10 annual or biennial mammograms, 1 in 2 women without breast cancer can expect to receive at least one false positive result.[22] How many women pay some kind of physical and psychological costs owing to false positives each year? In Germany, some 3 million to 4 million screening mammograms are performed annually. (Although German health insurance companies do not cover screening, German physicians often contrive reports of symptoms so that the companies will pay for screening mammograms.)[23] Out of these, some 300,000 come back as false positives. As a consequence of this large number of false positives, an estimated 100,000 women—none of them with breast cancer—are subjected to some form of invasive surgery every year. If the more than 50 million American women aged 40 or over followed the recommendation of the American Cancer Society to undergo annual screening, every year several million American women would get a false positive result. Many of these would also have their breasts biopsied. Estimates indicate that every year more than 300,000 American women who do not have breast cancer undergo a biopsy.[24]

Is there any way to stem this flood of false positives? Physicians in the Netherlands have attempted to do so by introducing more restrictive criteria for defining a positive mammogram. This policy has reduced the rate of false positives, but at the cost of increasing the rate of false negatives. In other words, more cancers are now missed during screening.

What can be reduced easily is the emotional impact of a false positive, that is, its psychological costs. Physicians could inform women about how frequent false positives are—for instance, by telling them that 1 in 2 women without cancer can expect to get one or more false positive results in a series of 10 mammograms. A woman who knows this will not be as shaken by a positive mammogram as a woman who does not. Nevertheless, I have met few women who have been informed by their physicians about how common false positives are.

False positives take a considerable toll on women's bodies and psyches.

About half of women who participate in regular screening are affected by this cost of mammography screening.

NONPROGRESSIVE BREAST CANCER

Eleanor was 49 years old when screening detected an early-stage breast cancer. Her surgeon removed part of her breast, and after radiation therapy, she finally got the good news that all cancer cells had been destroyed. Joyfully, Eleanor told all her friends that mammography and surgery had saved her and she could now live a carefree life. It is likely, however, that she might be mistaken and would have led an equally long and even happier life without having undergone the rigors of treatment.

She could have been in the second group of women who share the costs of mammography screening: They have a type of breast cancer that would never have been noticed during their lifetime but for the mammogram. There are two reasons for this phenomenon. First, there is a heterogeneous group of lesions called *ductal carcinoma in situ,* because the cancer is confined to the milk ducts of the breast and has not spread to the surrounding tissue. It cannot be detected by a breast exam, but it can be detected by mammography. Most breast cancers detected by mammography in women in their 30s, and about 40 percent in women in their 40s, are ductal carcinomas in situ. With higher age, the incidence decreases.[25] The clinical course of these cancers is not well understood, but it is thought that between 1 and 5 out of 10 ductal carcinomas eventually progress, becoming invasive cancer within 20 or 30 years. The other ductal carcinomas in situ seem never to spread at all and would not be noticed during a woman's lifetime except for the mammogram.[26] If a cancer is not invasive, neither the cancer itself nor treatment of it will affect how long the person lives. Physicians cannot predict which will become invasive and which will not. At present, almost every woman found to have a ductal carcinoma in situ is treated by lumpectomy or mastectomy. But half or more of these women would never have developed symptoms if left alone.

The second reason why a woman might have a cancer that would never have been detected in her lifetime is that some breast cancers, specifically if they are diagnosed in old age, progress so slowly that the women who have them will die before cancer can kill them. Like those women who have ductal carcinomas in situ that would not have proved fatal, these women typically undergo painful, traumatizing, and unnecessary therapy without the benefit of living any longer.

Most women are not aware that some cancers are nonprogressive. A survey of a random sample of American women revealed that few had heard about ductal carcinomas in situ, and 94 percent of the women surveyed doubted the existence of nonprogressive breast cancer.[27]

Women with nonprogressive or slow-progressive cancer pay a higher price for participating in mammography screening than women who receive false positives. The therapy they usually undergo as a result of having a positive mammogram lengthens not their lives, but the period during which they must live with a diagnosis of cancer and suffer through the treatment. For these women, early detection can *decrease* their quality of life.

RADIATION-INDUCED BREAST CANCER

A third group of women who bear some of the costs of screening are those who would not have developed breast cancer were it not for radiation exposure from mammography. The carcinogenic potential of radiation was recognized during the first decade of the twentieth century, when skin cancers of the hand were discovered among X-ray technicians. The earliest evidence of X ray–induced breast cancer was collected by German physicians in 1919, when chest X rays became a common method of diagnosing tuberculosis and other lung diseases.[28] The question of how large the risk is can be answered only indirectly—for example, based on the prevalence of breast cancer among women with tuberculosis who have frequent chest X rays or of various cancers among survivors of the nuclear attacks on Hiroshima and Nagasaki. Today there is general agreement that mammography can induce breast cancer, but researchers have come to different conclusions about how often this happens.[29]

The effect of radiation increases linearly with the dose: When the dose is halved, half as many women develop radiation-induced breast cancer; when the dose is doubled, this number also doubles, and so on. It does not seem to matter whether the total radiation dose is received in multiple exposures or in a single, brief exposure. The risk peaks 15 years to 20 years after the time of exposure. There is no evidence that radiation-induced breast cancer develops during the 10 years after the time of exposure or in women under the age of 25. The risk of radiation-induced breast cancer is lowest among women who start menstruating late, have their first child at an early age, breast-feed over a long period, and experience menopause early. Thus, the same hormonal factors that protect women from (nonradiation-induced) breast cancer seem also to decrease the risk of radiation-induced breast cancer.

The risk of radiation-induced breast cancer depends strongly on the age at exposure (Figure 5-1). For instance, when women in their 30s have a mammogram, the risk is twice as high as that for women in their 40s. The risk peaks around puberty and shows a sharp decline thereafter. During puberty, the usual source of radiation is not mammograms but thoracic and other types of X rays. For women over 60, the risk of radiation-induced breast cancer appears to be negligible, most likely because radiation-induced cancer typically takes up to two decades to develop.

According to current estimates, out of 10,000 women, between 2 and 4 who participate in mammography screening starting at age 40 will develop radiation-induced breast cancer, and one of them will die.[30] It is important to keep in mind that these figures are rough estimates that are based only on indirect evidence and can vary considerably with the imaging film used, the quality of the radiation, and other technical factors. For instance, the dose of radiation administered during a mammography in the early 1970s was about 10 times higher than that used today.

Mammography and other sources of radiation cause breast cancer in a small number of women. The best predictor of whether a woman will develop radiation-induced breast cancer is her age at exposure: the older she is, the lower the risk.

FALSE NEGATIVES

Besides these three groups of women who pay serious costs for mammography screening, there is another group who are not physically harmed but

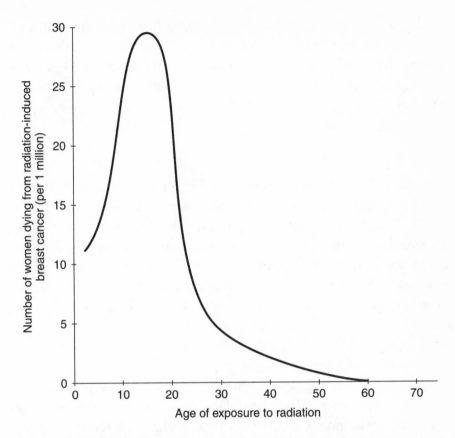

FIGURE 5-1. ***Deaths from radiation-induced breast cancer.*** The number of women who die from radiation-induced breast cancer depends strongly on age of exposure. (From Jung, 1998.)

are falsely assured. This group consists of women *with* cancer who receive a negative test result. The false negative rate is between 5 and 20 percent, with the higher rates in younger women. That is, out of 100 women with breast cancer, 80 to 95 will correctly get a positive result, but the others will be wrongly reassured by a negative result. False reassurance can result in forgoing the possibility for therapy, but this should not be counted as a cost of screening because not participating in screening would have had the same consequence.

A radiologist can try to decrease the number of false negatives, but at a price. A radiologist who wishes to minimize the possibility of missing a tumor has to classify more of the ambiguous results as suspicious (positive), thereby increasing the rate of false positives. There is a trade-off between false positives and false negatives. Radiologists who are highly accurate in detecting cancers (have few false negatives) typically have high rates of false positives.

FINANCIAL COSTS

Mammography screening is most cost-effective in woman aged 50 to 69 years. Screening women in this group biennially for 20 years results in costs of $21,000 per year of life saved.[31] For every $100 spent on screening, an additional $33 is spent following up on false positive results.[32] For breast cancer screening to benefit women who are not affluent or covered by generous health insurance, the quality of mammography will have to be improved or more cost-effective techniques will have to be discovered.

SUMMARY

Mammography screening can produce both costs and benefits for women. There are three major types of costs. First, every second woman who does not have breast cancer will have one or more false positive mammograms (in a series of 10 mammograms), and the resulting diagnostic follow-up can cause physical and psychological harm, such as the removal of breast tissue and increased anxiety. Second, a majority of women with a nonprogressive (or slow-progressive) breast cancer would never have noticed the existence of these abnormal cells during their lifetime except for the mammogram. In these cases, where the cells would never develop into invasive cancer, the treatment by lumpectomy, mastectomy, chemotherapy, and/ or radiation therapy with all their associated physical and psychological consequences are a second cost women have to bear. Finally, roughly 2 to 4 out of 10,000 women who do not have breast cancer will develop radiation-induced breast cancer because of mammography and one of them will die.

These are the main hazards of mammography screening from the point of view of the patient. I have not dealt here with the costs from the point of view of the physician. Physicians need to worry about protecting themselves against patients and lawyers who sue them for failing to detect an instance of cancer. This reality puts misses (false negatives) rather than false positives at the top of their worries, because lawyers tend to zero in on misses, not false positives. To decrease the possibility of false negatives, physicians tend to subject patients to batteries of tests—at the price of increasing false positives.

Before age 50, mammography does not seem to have benefits, only costs. Women at age 50, however, face the question of whether the potential benefits outweigh the costs. Each woman must decide for herself what the answer is. Her physician can help her understand what the benefits and costs are, but not how to weigh them. Her decision depends crucially on her goals, such as peace of mind, keeping her body unscarred, or a willingness to take (or not to take) the chance that she is one of the few who benefit from screening.

To choose wisely, women need to be informed about the risks. Are they?

What Are the Perceived Benefits of Screening?

Recall that mammography screening cannot decrease the incidence of breast cancer, only the mortality owing to breast cancer; no mortality reduction has been proven among women who participate in screening before age 50, and the mortality reduction from age 50 on is about 4 in 1,000 women.

INCIDENCE

A substantial proportion of women believe that screening actually decreases the incidence of breast cancer. These women mistake early detection for prevention. This confusion is fueled by leaflets from health organizations that emphasize the incidence of breast cancer, discussed later in this chapter. An organization that has an interest in getting as many women to participate in screening as possible has a conflict of interest when it attempts to inform women about the disease for which it offers screening. For instance, in a trial breast cancer screening program in the Swiss region of Morges (near Lake Geneva), women were informed about the screening and asked to participate.[33] Among these "informed" women, the proportion who mistakenly believed that screening would reduce their chance of developing breast cancer was higher than in the rest of Switzerland. As one Swiss colleague explained to me, "If we cannot get 70 to 80 percent of the women to participate, then the results of the study will be questionable." Therefore, health organizations are tempted to emphasize information that encourages women to participate rather than to correct their possible misunderstandings.

MORTALITY

The majority (55 percent) of more than 600 female patients who visited one urban medical practice in Chicago answered that mammography screening should start at age 30–35. It is unlikely that they got this mistaken belief from their physicians because, when questioned, none of the physicians in the practice stated that screening should start prior to age 40.[34] In a random sample of American women, some 40 percent believed that screening should start between the ages 18 and 39, and more than 80 percent believed that mammography had a proven benefit for women aged 40 to 49 years.[35] These women also believed that breast self-examination is even more beneficial than 10 years of annual mammograms.[36] These results indicate a striking degree of misinformation among American women. This is not to say that women in other countries are better informed; American women are just better studied.

How large is the perceived benefit of screening? In one study, 145 American women in their 40s were interviewed.[37] Their education and socioeconomic levels were above average—most had a college or postgraduate degree and a family income over $50,000. None of these women had had breast cancer. Nevertheless, consistent with the widespread and mistaken belief that screening before age 50 has proven benefits, more than 90 percent of them had had at least one mammogram. These women were asked:

Imagine 1,000 women exactly like you. Of these women, how many do you think would *die* from breast cancer *during the next 10 years* if they *are not* screened for breast cancer by mammography or a physician's exam?

and

Imagine 1,000 women exactly like you. Of these women, how many do you think would *die* from breast cancer *during the next 10 years* if they *are* screened every year or two for breast cancer by mammography or a physician's exam?

How many lives did these women think would be saved by 10 years of screening? Their average estimate—that is, the difference between their answers to the two questions—was 60 lives saved out of 1,000.[38] Recall that screening has *no* proven effect for women in their 40s, and across all ages, the benefit in a sample of 1,000 is only 1 life, not 60 lives, saved. A benefit of this size is illusory. Note that most of these well-educated women had had a mammogram. When they consented to screening, we have no reason to think that this was *informed* consent.

What Are the Perceived Costs of Screening?

In a survey of a random sample of American women, 92 percent said they believed that mammography cannot harm a woman who does not have breast cancer.[39] Of the rest, 3 percent referred to exposure to radiation as a potential hazard, 1 percent cited stress and anxiety, and an even smaller percentage raised the issue of false positives. Not one woman mentioned the unnecessary and traumatic consequences of a mammogram that detects a nonprogressive cancer.

Sources of the Mammography Illusion

The picture that emerges from these studies is this: Many women ascribe almost magical powers to mammography, and virtually none see harm in it. Where does this mammography illusion come from? People get their health information from three main sources: the media (including television, radio, newspapers, magazines, and the Internet), physicians, and leaflets from health organizations.[40] I will focus here on leaflets. How well do health organizations inform women about the costs and benefits of mammography screening? In 1997, Emma Slaytor and Jeanette Ward of the Central Sydney Area Health Service analyzed the information leaflets on mammography disseminated by all the cancer organizations, health departments, and mammography screening programs in Australia.[41] Does the content of the leaflets mirror what is and what is not in women's minds?

TABLE 5-2: *Information about mammography screening in 58 leaflets distributed by Australian health organizations.* For instance, in 60 percent of the leaflets the lifetime risk of developing breast cancer was mentioned. (Adapted from Slaytor and Ward, 1998.)

Information	How often provided
Lifetime risk of developing breast cancer	60%
Lifetime risk of dying from breast cancer	2%
Survival of breast cancer	5%
Relative risk reduction of death from breast cancer	22%
Absolute risk reduction of death from breast cancer	Never
Number that must be screened to avert one death from breast cancer	Never
Proportion of screened women who would be examined further	14%
False negative rate, or sensitivity	26%
False positive rate, or specificity	Never
Proportion of women with a positive mammogram who have breast cancer (positive predictive value)	Never

The information most frequently reported in the leaflets was breast cancer incidence (Table 5-2). Sixty percent of the leaflets specified a woman's lifetime risk of developing breast cancer (ranging from 1 in 11 to 1 in 16 in various leaflets). Two percent informed readers of a woman's lifetime risk of dying from breast cancer (which is lower). Although it is useful to know what the incidence of breast cancer is, this information does not reveal the costs or benefits of mammography screening. Screening cannot reduce incidence, only mortality. The leaflets' emphasis on incidence may be one reason for many women's belief that screening reduces incidence.

What did the leaflets say about mortality reduction? Only 22 percent reported mortality reduction at all. When this benefit was mentioned, it was invariably communicated in terms of relative risk reduction, which—as discussed earlier in this chapter—typically misleads laypeople into overestimating the benefits of screening. In other words, none of the leaflets reported benefits in terms of absolute risk reduction, number needed to treat, or other easily understandable ways.

The leaflets were uniformly silent about the possible costs of screening. For instance, not a single one mentioned the test's false positive rate despite the fact that about half the women undergoing repeated screening will experience one or more false positives. One-quarter of the leaflets mentioned that the test can lead to false negatives—for example, "mammograms pick up 90 percent of breast cancers" and are "not 100 percent accurate." Finally, how many of the leaflets informed women that only a few women who test positive in screening—1 out of 10 among 40-year-olds—actually do have breast cancer? Not one. The most striking feature of the leaflets is the information they left out.

This information policy does not seem to be peculiar to Australia.[42] The omissions—such as the possible costs of screening and the meaning of a positive test—correspond closely to the "blind spots" found in studies of American laypeople. The use of relative risks to report mortality reduction is consistent with women's overestimation of the possible benefits of screening. The omission of information about false positives is consistent with women's unnecessarily exaggerated anxiety about a positive result: A woman who has never heard that about 9 out of 10 women with a positive screening test do *not* have cancer will be unduly frightened when she has a positive result. As long as perceived benefits and costs are grossly distorted in the public mind, informed consent will remain out of reach when it comes to mammography screening.

Upon being asked why this is so, a psychologist who had interviewed hundreds of German women and found them ignorant about breast cancer and screening replied that many simply *do not want* to know any details about the disease or screening, and that this ignorance can almost be described as a collective defense mechanism.[43] Whatever the validity of this speculation, the analysis of the information leaflets summarized here demonstrates that the problem is not simply inside women's minds. There are those in the health information field who seem to prefer that potential participants in mammography screening not know too much.

From Innumeracy to Fear

It is impossible to weigh possible benefits against costs if one believes that there are no costs in the first place. Making an informed choice about breast cancer screening is further hindered by the specific anxiety surrounding breast cancer, fueled by the systematic exaggeration of its total incidence, of its incidence compared to other serious diseases, and of its incidence in young women. Of course, it is completely rational to fear developing a serious disease—what I am referring to is the added and unnecessary anxiety owing to misleading information provided by leaflets and the media.

"1 WOMAN IN 10"

In October 1999, the German weekly news magazine *Stern* ran a 13-page feature on breast cancer. In the subtitle, readers were told that every 10th woman develops breast cancer, a message repeated in the text. This was the only statistical information provided. The rest of the article played on readers' emotions through personal stories of hope and despair and the obligatory sensational photos (this time of a bevy of topless women wearing blue lace underwear and red boxing gloves—each of whom is missing a breast). The figure "1 woman in 10" (sometimes "1 woman in 9") has become a mantra in the popular press and breast cancer screening programs. This number has terrified many. But what does it mean?

What *Stern* did not report is that the 1-in-10 figure refers to a woman's cumulative chances of developing breast cancer *by the age of 85.* But most women die before then, and those who contract cancer at this advanced age will most likely die from some other cause. Table 5-3 explains how the

TABLE 5-3. *Risks of breast cancer and cardiovascular disease for 1,000 women, and the meaning of the "1-in-10" figure.* The data are based on the incidence and mortality rates reported in the Ontario Cancer Registry. (After Phillips et al., 1999.)

Age	Alive at beginning of interval	Incidents of breast cancer	Deaths from breast cancer	Deaths from cardiovascular causes	Deaths from other causes
0–9	1,000	0	0	0	7
10–19	993	0	0	0	2
20–29	991	0	0	0	3
30–34	988	1	0	0	2
35–39	986	3	0	0	3
40–44	983	5	1	1	4
45–49	977	8	2	1	6
50–54	968	11	3	2	11
55–59	952	12	3	5	15
60–64	929	12	3	9	25
65–69	892	14	4	16	36
70–74	836	13	5	28	51
75–79	752	11	6	52	70
80–84	624	9	6	89	95
≥85	434	5	7	224	203

1-in-10 figure plays out for a group of 1,000 women.[44] Four women developed breast cancer in their 30s and 13 during their 40s. By age 85, the total number of cases of breast cancer added up to 99 cases, which corresponds to the 1-in-10 figure. Thirty-three of these will have died of breast cancer. Using this natural frequency representation, we see that the majority of women with breast cancer did not die of the disease; about 3 in 100 women will die of it by age 85. And we also see that about six times as many women die of cardiovascular disease.

Of course, the numbers in Table 5-3 need to be adjusted for different populations. The goal here is simply to clarify the widely cited 1-in-10 figure. The tool that fosters insight into the 1-in-10 figure is the same as in Figure 4-2 (page 45): Start with a population of concrete cases, and break it down into subgroups. These frequencies are easily understood.

The 1-in-10 figure is rarely, if ever, explained in terms of frequencies as in Table 5-3. How do women interpret this statistic? Recall the group of well-educated American women aged 40 to 49 who had no history of breast cancer, each of whom was asked to imagine 1,000 women exactly like herself and to estimate how many of these women would die from breast cancer in the next 10 years. Their average estimate was 100, that is, exactly "1 woman in 10."[45] According to the authors of the study, this is a more than 20-fold overestimation of the actual risk of dying from breast cancer, consistent with Table 5-3. Among 1,000 45-year-olds (which is the average age of the women in this study), 5 died of breast cancer, not 100, within the next 10 years. Thus, on the assumption that most of these women had heard about the 1-in-10 figure, they apparently believed that it refers to the next 10 years rather than to the cumulative lifetime risk at age 85, not to mention mortality rather than incidence. Each of these misinterpretations fuels unnecessary levels of fear. Exaggerated fears of breast cancer may serve certain interest groups, but not the interests of women.

IS BREAST CANCER THE MOST DEADLY DISEASE FOR WOMEN?

Why does breast cancer arouse more anxiety than other threatening diseases? The leading cause of death among North American women is not cancer but cardiovascular disease, which attacks the heart and blood vessels (see Table 5-3). In a survey conducted by the American Heart Association, very few women (8 percent) reported knowing this fact. Similarly, in a survey by The National Council on Aging only 9 percent of women said the disease they most feared was heart disease compared with 61 percent who most feared cancer, predominantly breast cancer. Even among cancers, lung cancer—not breast cancer—is the leading cause of death among American women (Figure 5-2). Only 25 percent of women were aware of this fact.[46]

In the United States, prostate cancer is more frequent than breast cancer (Figure 5-2). It also takes almost as many lives. But there is no anxiety among men about prostate cancer comparable to women's anxiety about breast cancer. Interestingly, unlike breast cancer, prostate cancer is presented in the media as an old person's disease. However, its incidence, mortality rate, and the mean age at diagnosis are in fact very similar to those of breast cancer.[47] Colorectal cancer is the third most deadly cancer among women (after lung and breast cancer). "Colorectal cancer" is an umbrella term for colon cancer and rectal cancer, colon cancer being the more common of the two. Like lung cancer, colorectal cancer rarely makes it into the headlines or onto the front pages. Consistent with what the media pres-

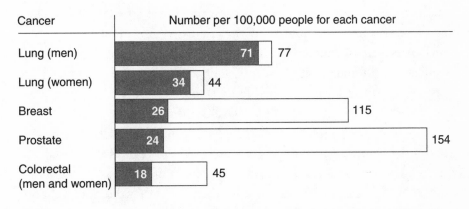

FIGURE 5-2. *Mortality (dark area) and incidence (dark plus white area) of the four most frequent cancers in the United States.* For instance, out of 100,000 men, 77 were diagnosed every year of lung cancer and 71 died from it (1990–1995, caucasians only). (Based on the figures reported by Wingo et al., 1998).

ents, relative to breast cancer the public tends to underestimate the risk of cardiovascular disease and other cancers. A defender of the media might retort that it is public perception that the media coverage follows, not vice versa. There need be no contradiction—each probably follows the other.

TARGET THE YOUNG

In a 1991 issue featuring an article on breast cancer, *Time* ran on its front cover a picture of a young woman with highlighted, bare breasts. In the years since then, many other magazines have followed suit, thus creating the impression that young women have the most to fear from breast cancer.[48] In one study, some 85 percent of the case stories and anecdotes about breast cancer presented in magazines such as *Glamour, Vogue, Scientific American, Time,* and *Reader's Digest* were found to be about women under 50.[49] When the breast cancer incidence rose dramatically in the 1980s and early 1990s, popular magazines portrayed this increase as a mysterious epidemic targeting young, liberated, professional women.[50]

Age is indeed the highest risk factor for breast cancer—but it is the old, not the young, as suggested by popular magazines, who are most at risk. The average age at diagnosis is around 65 years (see Table 5-3). Moreover, the increase in incidence has not been observed in younger women, and the largest upsurge has been among women over 60. Most, if not all, of this increase can be attributed to the increased use of mammography, which detects many cancers that may never become clinically relevant (such as ductal carcinoma in situ).[51] Consistent with this view, mortality rates from breast cancer have remained fairly stable as incidence has increased. The incidence of breast cancer finally leveled off in the early 1990s, after the "bubble" of increased screening had passed through the health system.

What has remained relatively constant over the years is the prevalence among men, who account for about 0.5 percent of all breast cancer diagnoses.[52] Men do not participate in screening. But if they did, we might observe an increase in incidence among them as well.

Do women's beliefs mirror the emphasis in the media on young victims of breast cancer? Some 700 female patients at a Chicago academic internal medicine practice staffed by 31 physicians were asked whether breast cancer is more common among women aged 65 or women aged 40. Only 28 percent knew that the prevalence is higher among 65-year-olds than 40-year-olds.[53] Similarly, a study of women in North Carolina found that 80 percent did not know that older women have a higher prevalence of breast cancer.[54] Yet in the Illinois study, all the physicians in the practice correctly answered that breast cancer risk increases with age. This finding suggests that the patients did not get this misconception from their physicians, though the physicians also failed to correct it. The media focus on the threat that breast cancer poses to the young is badly misdirected. Older women have the most grounds for concern.

From Innumeracy to Prophylactic Mastectomy

A combination of innumeracy and fear can contribute to women allowing themselves to be subjected to treatments that they would otherwise reject as needlessly alarmist or even dangerous in themselves. The psychologist Robyn Dawes reported on a dramatic example involving a Michigan surgeon whom a newspaper hailed as a pioneer in the treatment of breast cancer.[55] The surgeon urged all women over 30 to have an annual mammogram. Moreover, he advocated removing the breasts of healthy women and replacing them with silicone implants. If you cannot follow his argument justifying this practice, don't worry—but be wary.

The surgeon argued that 57 percent of women in the general population are in a group at high risk of developing breast cancer, and that 92 percent of all breast cancers are found in this group. Furthermore, he claimed that 1 in 13 women in the general population (whether at high or low risk) develops breast cancer between the ages of 40 and 59. From this he concluded that 1 in every 2 or 3 women in the high-risk group will develop breast cancer between the ages of 40 and 59.

On the basis of these "estimates," the surgeon recommended that women without breast cancer who are in the high-risk group—that is, the majority of women—have prophylactic (preventive) mastectomies. This measure, he argued, would save them from having to face the risk of cancer and its consequences, including the possibility of death. Over the course of two years, he removed the "high-risk" breasts of 90 cancer-free women and replaced them with silicone implants.

Persuaded by the surgeon's argument, these women may have believed they were sacrificing their breasts in a heroic exchange for the certainty of saving their lives and protecting their loved ones from suffering and loss. None of them, and none of their loved ones, seem to have questioned the surgeon's numbers and reasoning.

To check whether the surgeon's reasoning is sound, let us take a minute to draw a tree like that in Figure 5-3. Think of 1,000 women. According to the surgeon, 570 (57 percent) of them belong to the high-risk group. According to him, 77 of the 1,000 women (that is, 1 in 13) will develop breast cancer between the ages of 40 and 59, of whom 71 (that is, 92 percent) are in the high-risk group. Thus, 71 out of the 570 high-risk women will develop breast cancer, which is about 1 in 8—not 1 in 2 or 1 in 3 as the surgeon concluded.

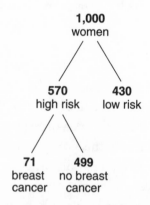

FIGURE 5-3. *A frequency representation of the risks communicated by a Michigan surgeon to his clients.* The surgeon wrongly concluded that 1 out of 2 or 3 women in the high risk group will develop breast cancer. When we transform his percentages into frequencies, we can easily see that this number is about 1 in 8 (71 in 570), not 1 in 2 or 3. When, in addition, we use realistic figures instead of his subjective ones, this number is about 1 in 17.

Now we can see the flaw in the surgeon's reasoning: He drew the wrong conclusions from the risks he specified. Just as alarming, the numbers he used in his computations are themselves inflated (1 in 13 is unrealistically high for women between 40 and 59). In Table 5-3, we can see that in a group of 1,000 women, 36 women will develop breast cancer between the ages of 40 and 59. From this more realistic figure, we would expect that only 1 in 17 high-risk women will develop breast cancer between the ages of 40 and 59.[56] This result indicates that out of the 90 women on whom the surgeon performed prophylactic mastectomy, some 85 would not have developed breast cancer anyway. (And several of the others could most likely have been treated less radically—for instance, by a lumpectomy.) In this case, the joint innumeracy of the surgeon and his patients had tragic costs—for the patients.

What is the actual benefit of prophylactic mastectomy? A recent study followed up 639 women with a family history of breast cancer who had undergone prophylactic mastectomy of both breasts at the Mayo Clinic in Minnesota.[57] The women were classified as either high risk (for example, women with a mutation in the breast cancer genes BRCA1 and BRCA2; women with one or more first-degree relatives with breast cancer) or moderate risk; their median age at mastectomy was 42 years and the median follow-up time was 14 years. The result is shown in Table 5-4 (page 85).

Ways to Present the Benefit

High-risk group

Absolute risk reduction. Prophylactic mastectomy reduces the number of women who die from breast cancer from 5 to 1 in 100. That is, the absolute risk reduction is 4 women in 100 (4 percent).

Relative risk reduction. Prophylactic mastectomy reduces the risk of dying from breast cancer by 80 percent (because 4 saved out of 5 is 80 percent). Remember that the relative risk reduction is the absolute risk reduction (4 in 100) divided by the proportion of patients who die without treatment (5 in 100).

Number needed to treat. The number of women who need to undergo prophylactic mastectomy to save one life is 25 because 4 in 100 (or 1 in 25) deaths are prevented by prophylactic masctectomy.

Moderate-risk group

Absolute risk reduction. Prophylactic mastectomy reduces the number of women who die from breast cancer from 2.4 to 0 in 100. That is, the absolute risk reduction is 2.4 women in 100 (2.4 percent).

Relative risk reduction. Prophylactic mastectomy reduces the risk of dying from breast cancer by 100 percent.

Number needed to treat. The number of women who need to undergo prophylactic mastectomy to save one life is 42.

All three presentations of the result are correct, but can suggest different amounts of benefits and evoke different emotional reactions in women. For instance, for high-risk women, prophylactic mammography reduced the risk of dying from breast cancer by 80 percent, or by 4 percent, depending on whether the benefit is presented as a relative or an absolute risk reduction. Moreover, in terms of the number needed to treat, out of every 25 women at high risk, the life of one was saved, whereas the other 24 had no benefit from mastectomy (because most high-risk women do not die of breast cancer, even if they keep their breasts, and a few die of breast cancer even after they had both breasts removed). In the moderate risk group, one life was saved for every 42 women who underwent treatment, which also means that 41 out of every 42 women lost their breasts without a benefit.

Women who are contemplating prophylactic mastectomy should know these numbers in order to be able to make an informed decision. Furthermore, it is essential that they understand what the numbers mean, that is, the difference between the absolute risk reduction, the relative risk reduction, and the number needed to treat. As the cases of the 639 women at the Mayo Clinic show, prophylactic mastectomy can save lives. At the same time, however, it does not provide absolute certainty because seven women got breast cancer even after a prophylactic mastectomy, and for the vast

TABLE 5-4: *Reduction in breast cancer mortality resulting from prophylactic mastectomy* (Hartmann et al., 1999).

Treatment	Deaths (per 100 women)	
	High-risk group	Moderate-risk group
Prophylactic mastectomy	1	0
Control (no mastectomy)	5	2.4

majority of women treated, the treatment led to a loss in quality of life without a prolongation of life.

Conclusions

Breast cancer is a fairly unpredictable disease, with some tumors growing quickly and others so slowly that they never would cause any symptoms; and there are yet others in-between. In making a decision about breast cancer screening, women face a situation in which the illusion of certainty and all three kinds of innumeracy interact. The illusion of certainty is fostered by physicians who routinely present patients with a choice between certainty and risk, although the real choice is always between two risks, namely, the risks of screening and those of not screening. The illusion of certainty is fueled by leaflets distributed by health organizations that report the benefits of screening without mentioning its hazards. Ignorance of risks seems to be the rule rather than the exception: for instance, the massive number of false positives is rarely remarked upon, nor is the existence of nonprogressive cancers where early detection has no benefits, only costs. Furthermore, benefits are often communicated in terms of relative risk reduction, which impress and mislead laypeople. As a result, knowledge of the benefits and costs among women is disturbingly low, and informed consent is hardly possible. Finally, the accuracy of mammography is often communicated in terms of probabilities, which befuddles the minds of most physicians—not to mention their patients.

Of course, innumeracy is only one factor perpetuating this deplorable state of affairs in modern medicine. There are also institutional factors such as physicians' attempts to avoid being sued, conflicts of interest within health organizations that want both to inform patients and to get them to participate in screening, and—last but not least—emotional reasons such as many patients' preference for reassurance over realistic information.

Of the various means capable of changing this state of affairs—legal, professional, and other—the tools for combating innumeracy are the most cost-effective and easiest to use. These tools include transparent methods of risk communication and mental defogging. Once insight has replaced the innumeracy that now prevails, calls for reform may bring about the necessary institutional and professional changes.

Perform all [these duties] calmly and adroitly, concealing most things from the patient while you are attending to him. Give necessary orders with cheerfulness and serenity, turning his attention away from what is being done to him; sometimes reprove sharply and emphatically, and sometimes comfort with solicitude and attention, revealing nothing of the patient's future or present condition.

Hippocrates

6

(UN)INFORMED CONSENT

The nineteenth century saw a struggle among three different visions of the physician: the artist, the statistician, and the determinist. The French medical professor Risueño d'Amador, for one, promoted the vision of the physician as an artist who relies on medical "tact" and his intuitions about the individual patient. His rival Pierre-Charles-Alexandre Louis, in contrast, had little respect for medical tact and wanted instead to see the evidence. Louis became famous for rejecting the established doctrine of bloodletting as a medical treatment. By collecting data, he showed that slightly more people who were bled died than people who were not, concluding: "We shall hear no more of medical tact, of a kind of divining power of physicians."[1]

At that time, using statistics to test the effectiveness of medical practices was revolutionary. The idea was inspired by the statistical methods of Pierre-Simon Laplace in astronomy and Adolphe Quetelet in social science. But statistical evidence was frowned upon not only by medical "artists." The French physiologist Claude Bernard rejected the images of both the physician-artist and the physician-statistician. For Bernard, science meant certainty. He ridiculed the use of statistical information:

> A great surgeon performs operations for [kidney] stone by a single method; later he makes a statistical summary of deaths and recoveries, and he concludes from these statistics that the mortality law for this operation is two out of five. Well, I say that this ratio means literally nothing scientifically and gives us no certainty in performing the next operation.[2]

Averages were, for Bernard, no substitute for the laws that determine each individual case, and a true determinist would settle for nothing less. The way to discover these laws, in this view, was by experimentation, not the use of statistics. In the nineteenth century, statistical data were still considered antithetical to the scientific method. Whereas science was about certainty, statistics was about uncertainty; therefore, statistics was not a proper scientific tool. The German-Hungarian physician Ignaz Semmelweis's statistical studies of childbed fever and scurvy are as legendary as the reluctance of the relevant authorities to enact the preventive measures that his statistics suggested. Unlike in physics, statistical thinking emerged slowly in medical diagnosis and treatment.

The distinction that Bernard drew between statistics and experimentation was finally closed in the 1920s and 1930s, when the English statistician Sir Ronald Fisher united statistics and experiment into what he called the "scientific method." The medical statistician Austin Bradford Hill pioneered the application of Fisher's randomized control experiments to randomized trials in medicine and was knighted for this achievement in 1961. Praised for its "concern for the welfare of the individual," his work reconciled medical statistics with experimentation, thereby reconciling the aggregate with the individual.[3]

These visions of the physician have shaped today's conflicting attitudes about who is entitled to make medical decisions. The physician? The patient? Both? For physicians who see themselves as artistic virtuosi of medical tact, the patient is an uninitiated audience who may applaud, but not participate in, decision making. After all, a conductor would not ask his audience for advice on how to play Beethoven. Consistent with the image of the physician-artist, doctors made virtually all the decisions; patients felt reassured and did what they were told. The patient's body was treated as if it were the property of the doctor, who decided whether it was to be given drugs or subjected to surgery. Some physician-artists did not even allow patients to look at their own medical records.[4]

Today, medical decision making is changing; patients have become more and more involved. This change was marked by the publication of the book *The Silent World of Doctor and Patient* by a Yale physician, Jay Katz. When the book appeared in 1984, surgeons almost uniformly attacked Katz's view that patients should have a say in what happens to their bodies. The magnificent vision of the physician as an artist was at odds with the idea of an informed and mature patient.

Bernard's deterministic ideal has left a different mark on medical decision making. Physicians who believe in this ideal see medical decisions as choices between certainty and risk. The book *Medical Choices, Medical Chances*, written by a team led by Dr. Harold Bursztajn of Harvard Medical School, however, has been influential in making clear that medical choices are almost never between certainty and risk, but rather between two risks. Tests and treatments are often inconclusive and can have side effects, and certainty is usually out of reach.

The book opens with the case of a 21-month-old boy who was admitted to a leading American teaching hospital. The boy, who had an ear infection, was pale, withdrawn, and severely underweight. Though starving, he often refused to eat. The team of well-meaning doctors who treated the boy believed that its responsibility was to identify the cause of his illness with "certainty." Although they considered all actions not directed toward this goal to be risky, they saw no serious risk in attempting to reach certainty through relentless testing, which required repeatedly drawing blood from the emaciated child. Once the diagnostic machinery was turned on, distinguished specialists performed numerous biopsies, six spinal taps, and a host of further tests—many of them designed to diagnose what were then untreatable diseases. The doctors felt they couldn't take the chance of failing to identify the cause of the boy's illness. What did the tests reveal?

Nothing certain. Between each invasive test and the next, however, the little boy refused food more and more often, and after six weeks of a struggle for certainty, he died.[5] In an uncertain world, certainty can be a dangerous ideal.

Louis's vision of the physician who bases diagnostic and treatment decisions on statistics, on the other hand, lends itself to the development of a type of doctor-patient interaction in which both parties can discuss what to do on the basis of the available evidence and the patient's preferences. The modern descendent of this vision has been termed *evidence-based medicine*. Fortunately, an increasing number of physicians practice evidence-based medicine. That is, they base their diagnostic and treatment decisions on the available evidence rather than on local clinical procedure and personal preference. Ideally, the physician and the patient decide on medical care together, with the physician being the expert on possible treatments and the patient the expert on what she or he wants and needs.

Unfortunately, real medical decision making often fails to live up to this laudable ideal. It is telling that the term "evidence-based medicine" had to be coined at all—think about a group of natural scientists in need of promoting evidence-based physics. A representative of the World Health Organization recently estimated that only 40 percent of medical practitioners in the United States choose treatments for their patients on the basis of scientific evidence; the others use treatments that are possibly ineffective. One reason for the reluctance in medicine to rely on evidence—which persists today, more than 100 years after Bernard—is that many physicians still have difficulties drawing diagnostic inferences from statistics.

Informed Consent

What do physicians think about informed consent and shared decision making? Below, the second part of the discussion among the 60 physicians, introduced in Chapter 2, shows that their opinions differ radically.

President: Medicine today still resembles the church in the sixteenth century. In surgery, it's mostly men, they speak a strange language, and their clients are reassured after confession. The men wear strange wardrobes and perform ritualistic operations on the appendix. What we need is a reformation. Martin Luther brought the Bible to the people by translating it from Latin into German. We have to bring the evidence to the patients by translating relative risks and other confusing language into natural frequencies and transparent language.

Organizer: Let me expand. A main target of criticism for the Reformation was the buying of indulgences, where people were made to believe that they could buy God's absolution of every sin. This is what priests told them, and people believed it because they had no access to the Bible itself. Today's analogy is the medical business, where patients are made to believe that they can buy a cure for every disease. And there is a second parallel: new information technology. The printing press was the vehicle of the Reformation; it spread the word translated by Luther to the people. For the first time, everyone could read the text rather than depend on what a priest said. Today we have a second revolution in information technology: the Internet. The Internet allows access to medical information that was difficult to obtain before. Noncommercial groups of physicians, such as www.cochrane.org, distribute the information patients need to know over the Internet.[6] Luther leveled the disparity between priest and layperson, the initiated and the ignorant. The Internet can help us to level the disparity between the physician and the patient, the infallible and the uninformed. This is my vision of a reformation: Have doctors making decisions about treatments use the best available evidence and consider the patients' goals.

Gynecologist: Well, I agree with this vision, but my reality is different. I tell women about costs and benefits so that they can make up their minds. But few are interested in the numbers—most decide irrationally. This woman's neighbor had breast cancer—this is why she goes to be screened.

Professor M: But we do have a responsibility to inform patients. Women are badly informed about mammography screening. Many physicians just invoke feelings of guilt to make them participate. "I hope you finally went. You still didn't?" Many women even believe that mammography could prevent breast cancer, [the way] brushing your teeth prevents cavities. Feminist groups have fought for screening. But few feminists have been informed that early detection is not always a benefit—for instance, when a slow-growing cancer is detected that never would have progressed into an invasive one. The minimum a patient needs to be told is the goal of screening, how often false positives and false negatives occur, the benefits and costs of mammography, and the financial consequences. A physician must reveal to a woman that he will earn money from performing a mammogram, but not when she declines it.

Dr. B: But patients do not like the prospect that screening can also be harmful. There are those who, in despair, fixate on mammography and think it's their rescue from breast cancer. It's their hope.

Professor M: Only 1 out of 1,000 women benefits from 10 years of screening. In other words, 99.9 percent who participate in screening have no benefits, only potential harm. However, this is not the complete story. Swedish studies found that the total number of deaths—from breast cancer or some other cause—is the same for women who are screened and those who are not.

Dr. B: Why is this?

Dr. C: Maybe from car accidents on the way to screening.

Professor M: We don't know. And despite excellent studies on mammography screening, we continue to live with uncertainty.

Organizer: We must learn to admit that there are different opinions, that there is no final answer. The question is: Go to screening and run the risk of possible costs, such as an early detection and treatment of a cancer without reducing mortality, resulting in a loss of quality of life? Or not go, betting that I am among the 999 out of 1,000 who will not benefit from screening, as Professor M pointed out, but run the risk that I am actually the one woman whose life would be saved?

Dr. A: Physicians in Essen, Germany, amputated one or both breasts of some 300 women, despite most of them not having cancer. When this was proven, one physician set fire to his records and then himself. A Swedish study resulted in 4,000 unnecessary breast amputations. Is all this worth the 1 in 1,000 benefit?

Dr. C: Why so many unnecessary breast amputations?

Dr. A: In the German case, shoddy diagnoses. But in general, the diagnoses of histologists are consistent with one another in only 70 percent of cases.

(Breast cancer specialist shakes head in disagreement.)

Organizer: The pressure on women to participate in mammography screening is enormous. "Come back after six months, you should have a follow up mammogram . . ."

Dr. D: But why do you insist on informing? Most patients do not want to be informed. It's all psychology. They are anxious; they fear the worst; they want to be reassured. They do not want to see numbers.

Gynecologist: After a mammogram, it is me, the physician, who is reassured. I fear *not* recommending a mammogram to a woman who may later come back with breast cancer and ask me "Why didn't you do a mammogram?" So I recommend that each of my patients be screened. Yet I believe mammography screening should not be recommended. But I have no choice. I think this medical system is perfidious, and it makes me nervous.

Dr. A: Do you yourself participate in mammography screening?

Gynecologist: No, I don't.

Organizer (addressing the audience):   I would like to know how many of you participate in mammography screening. For men, the question is "If you were a woman, would you participate?"

Organizer (seeing that no one has raised his or her hand):   Hmm. How many of you do not participate, or would not participate? How many are undecided?

Organizer (after counting hands):   No one here participates herself in screening. Fifty-five say they do not participate. Five are undecided—men, who haven't thought about it.

President: We need to put hard evidence into patients' minds and let the illusions out. At the same time, we need to take their anxieties and their need for rituals seriously. Every doctor practices a bit of voodoo and mysticism, even the star surgeon. Patients expect this. But most important, we need to learn to use our brains and finally enter the age of Enlightenment, which Kant asked us to do long ago. If not perfectly rational, then at least enlightened.

Dr. F: Informed consent is not just prevented by the anxious patient who refuses to think. Many physicians do not understand the risks in the first place, and this may affect women's emotions and anxieties.

Breast cancer specialist: I admit that until a few years ago, I emotionally overestimated the power of technology. In my field, doctors focus only on not overlooking a cancer. When a false positive is revealed after cutting into a breast, the patient is happy, and the doctor is too. The patient does not ask, "Why did you cut in the first place? Why not a needle biopsy or other minimally invasive diagnostics?"

Organizer: The patient says to herself, "Thank God that they looked so carefully, and now everything is OK." Invasive surgery with a benign result relieves the patient and makes her grateful to the physician.

Decision researcher: An AIDS counselor once told me that there were never any false positives. I asked him, "Would you actually notice a false positive?" That made him pause, and he thought for a while and remarked, "Actually, most likely I would not." Similarly, radiologists do not follow up on patients, do not keep statistics, and do not conduct quality control studies. They don't keep track of which positives turn out to be false positives. In addition, to talk with physicians about false positives often evokes anxiety and a defensive attitude on their part.

Dr. A: I sit through many a seminar for continuing education for physicians, where we explain at length what "sensitivity" [true positive rate] and "specificity" [true negative rate] mean. Then we ask a participant what the probability of a positive test is called given the presence of a disease. One physician answers "specificity." I say, not exactly, try again. "Oh," he corrects himself; "I wanted to say '1 minus specificity.'" I say, try again.

President: Few doctors are trained to judge and evaluate a scientific study. I myself chose to be trained as a surgeon in order to avoid two things: statistics and psychology. Now I realize they are both indispensable.

The discussion reveals different views on informed consent and shared decision making. One group of physicians perceive patients as emotional beings who want reassurance rather than information. Others emphasize that physicians have a moral obligation to inform their patients, and that the lack of informed decision making is due not just to emotionally or intellectually handicapped patients, but also to physicians' faults, including their innumeracy. Finally, as Professor M points out, informed consent must also entail the physician's revealing when his or her own costs and benefits concerning a treatment differ from those of the patient. This situation is illustrated clearly in the disturbing fact that physicians recommend screening to their patients although they do not participate themselves.

PHYSICIAN AND PATIENT

Is it "natural" that patients are expected to trust their doctors' judgment without question, that they have little say in medical decision making and are badly informed? Although the omnipotent role that physicians play in modern society has been defended as a long-standing tradition, there is nothing natural about this form of doctor-patient relationship. For instance, records in Bologna dating from the late sixteenth century through the seventeenth century reveal a strikingly different doctor-patient relationship. Patients were expected to pay only if they were healed.[7] The practitioners were contractually bound by "an agreement of cure" to heal the patient within a specified time and for a specified sum. This "horizontal" as opposed to "hierarchical" model of the physician-patient relationship empowered patients. There was a tribunal before which patients could sue the practitioners, licensed or unlicensed, when they broke the cure contract. By the end of the eighteenth century, however, patients no longer received a promise of cure but instead a promise of orthodox treatment and protection. Payment according to results more or less disappeared with the professionalization of modern medicine. The status of American physicians today—known as "allopaths"—reflects the outcome of a century-long struggle among competing groups of healers including homeopaths, midwives, barber-surgeons, and wise women. Well into the nineteenth century, American physicians were little respected and were underpaid. The twentieth century witnessed the emergence of a powerful medical monopoly and the birth of a highly respected profession. Jay Katz sees a link between physicians' historical struggle for dominance and their relationship with patients today: "Physicians' quest for political power mirrors the quest for interpersonal domination of the physician-patient relationship."[8]

The dominance of physicians and submission of patients in modern medicine began to change when, on October 22, 1957, Justice Absalom F. Bray of the California Court of Appeals coined the term "informed consent." A man with paralyzed legs had sued his physician for failing to warn him of the risk of paralysis inherent in the treatment for limping that he had undergone. At the end of his opinion, Justice Bray wrote: "In discussing the element of risk a certain amount of discretion must be employed consistent with the full disclosure of facts necessary to an informed consent."[9] His opinion marked the beginning of the legal history of informed consent. The fact that Bray left room for "discretion" as well as "informed consent" reflects a tension that still exists today. Informed consent refers to an ideal practice in which the physician informs the patient of the risks associated with a treatment, including its side effects, on the basis of which the physician and the patient jointly decide what to do. Informed consent is more appropriately called "shared decision making." The two parties bring different knowledge to the decision—the physician knows about diagnostic tools and treatment options, while the patient knows what his or her goals and values are. For instance, a woman might or might not prefer to keep her reproductive organs intact at the risk of lowering her life expectancy. Expressed in general terms, a patient might judge the possible benefits of a treatment as worth its costs, or the patient might find the costs too high and the benefits too small to be worth the trouble.

IPHIGENIA

Consider the case of a woman whom Katz called Iphigenia (to protect her anonymity)—after the classical Greek playwright Euripides' Iphigenia, who was saved by the goddess Artemis from sacrifice by her father. The case illustrates how physicians and patients can make the transition from a relationship founded on authority to one based on informed consent.[10] Iphigenia, a 21-year-old single woman, discovered a lump in her breast, which a biopsy revealed to be a malignant lesion. She agreed to a mastectomy, the extent of which her surgeon would determine by what he discovered during the operation. The surgeon had not explained the pros and cons of alternative therapies to Iphigenia because he firmly believed that they all were inferior to surgery. As the day of the operation approached, however, he began to doubt the wisdom of proceeding without giving her details about alternatives, particularly radiation therapy. On the evening before the operation, he went to the hospital and finally told her about his concerns about remaining silent. After they talked for a long time, Iphigenia decided to postpone the operation and eventually opted for a lumpectomy (excision of the tumor only) followed by radiation therapy. At a panel discussion on the treatment of breast cancer some time later, Iphigenia spoke knowledgeably about why she had made this decision and movingly expressed her joy at being able to start her impending marriage physically intact. The discussion was heated. The physicians defended their preferred treatment and attacked those defended by their colleagues—just as financial advisors tend to favor their own methods. Despite the disagreements, however, the physicians were fairly united in the belief that it was irrational of Iphigenia's surgeon to allow her to make the decision whether to have surgery. Even physicians have trouble deciding which treatment is best, they reasoned, so how can a physician allow a patient to decide?

NANCY REAGAN

Iphigenia decided for a lumpectomy over a mastectomy. Faced with early-stage breast cancer, former First Lady Nancy Reagan weighed the pros and cons differently and decided to have a radical mastectomy. In her words:

> At the time of my operation, there were some people, including doctors, who thought I had taken too drastic a step in choosing the mastectomy instead of a lumpectomy, which involved removing only the tumor itself and a small amount of tissue—but also weeks of radiation. I resented these statements, and I still do. This is a very personal decision, one that each woman must make for herself. This was my choice, and I don't believe I should have been criticized for it. For some women, it would have been wrong, but for me it was right. Per-

haps, if I had been 20 years old and unmarried, I would have made a different decision. But I've already had my children and I have a wonderful, understanding husband.[11]

Not only can values differ from person to person, but they can change over the course of a lifetime. As soon as physicians agree to share information and responsibility, patients become more knowledgeable about treatments and physicians more aware of patients' values. To use an analogy, the question of how to build a power plant is a technical one best handled by engineers. The question of whether and where a power plant should be built, however, is only partly technical; it is also a social and political question that should not be decided solely by engineers.

Uninformed Consent

Informed consent is a beautiful ideal, but why is it still so rare? The silent world between doctor and patient is one reason: Patients often do not know what questions to ask. The emotional world of the patient is another: Many do not dare taking an active part in the decision process. And then there is innumeracy among physicians, which puts informed consent out of reach. The fundamental building block of the uninformed, immature patient, however, is a world of illusionary certainty.

THE ILLUSION OF CERTAINTY

Some physicians argue that disclosing the uncertainties of diagnosis and treatment to patients would be counterproductive because the patients would not understand them, would not want to know the numbers or want to hear that treatments have potential hazards as well as benefits—and they might seek out a physician who offered them certainty instead. In this view, the physician's main task is to reassure the patient. According to the physician Jay Katz, until recently physicians have assumed an attitude of certainty toward their patients, despite a wide variety of competing treatments for the same ailment. Without the tools of controlled experiments and statistics, it was difficult to evaluate the claims of various thera-pies.[12] Every proposed treatment, however reasonable or absurd, found its enthusiastic adherents. These included surgery, a variety of medications (mineral, vegetable, and animal), dieting, bleeding, purging, exorcism, in-ducing perspiration, the laying on of Queen Elizabeth I's royal hands, and applying goat dung. When one treatment fell into disrepute, it was quickly replaced by another one, which was then touted as the single best one. Even in medieval times, surgeons removed part or all of diseased breasts, and leading surgeons accused colleagues who did not remove every bit of tissue for being too timid in a debate that is still going on today.

Many a physician confronts the patient with an apparent choice between *certainty* and *risk* rather than a choice between risks. Each alternative car-ries its own uncertain consequences, which need to be compared for an in-formed decision to be made. Consider, for instance, Dr. Standing's statement that he would not inform a patient of evidence about errors in mammography and would perform a biopsy on a patient with a positive mammogram in any case. This course may sound harmlessly prudent, but for the patient a biopsy carries costs. In an article in *The New Yorker* enti-tled "Whose Body Is It, Anyway?" the surgeon Atul Gawande related the case of a woman in her 40s whose annual mammogram revealed an area of "suspicious" calcification in her left breast on three separate occasions.[13] Each time, a surgeon took her to the operating room and removed the tis-sue in question. Each time, the tissue proved to be benign. Now the woman is again confronted with a suspicious mammogram of the same breast, and the doctor tells her that she ought to have a biopsy to exclude the possibil-ity of malignancy. She already has three raised scars scattered across the breast, one of them almost 3 inches long, and so much tissue has been re-moved that her left breast is markedly smaller than her right one. Should she really have another biopsy?

This woman's choice is between two risks, not between certainty and a risk. To choose, she needs to know the costs and benefits of each option and to evaluate these in light of her own goals, which may differ from her physician's. And she may have to live with a residual uncertainty, because not everything is known as precisely as we would wish. The lifeblood of the illusion of certainty, however, is thinking in black-and-white terms rather than in shades of gray: "Either my mammogram will be normal, and I do not have to worry about breast cancer, or it will not be and I will die an agonizing, horrible death."[14] Neither of these two outcomes is truly certain.

DO PATIENTS ASK QUESTIONS?

By definition, informed consent requires an informed patient, not just a consenting one. However, many patients are poorly informed about risks—and not only concerning breast cancer. In one study, patients in the waiting rooms of a Colorado clinic serving army personnel and an Oklahoma clinic serving a low-income urban population were asked about the standard tests for diagnosing common illness and other diseases such as strep throat infection, HIV, and acute myocardial infarction.[15] Each patient judged (1) the probability that a person such as himself has the disease before being tested, (2) the sensitivity of the test, (3) the specificity of the test, and (4) the probability of his having the disease given a positive test result. Most patients estimated these four probabilities to be essentially the same for all the diseases—whether the disease was rare or not and whether the test was accurate or not. To find out if patients' ignorance stemmed from the fact that they had no experience with these diseases, for each illness or disease the authors looked at only those patients who had been tested or treated for it or had attended an office visit with a family member or friend who had been tested or treated for it. In the Oklahoma clinic, the estimates of patients who had experience with the illness or disease were only slightly more accurate than other patients' estimates, and in the Colorado clinic, they were no more accurate. Even the experienced patients made very inaccurate judgments, which suggests that their physicians had never explained the risks to them or had done so in a way that was hard to understand or at least hard to remember.

Do patients ask for information concerning risks that doctors fail to deliver, or do patients rarely ask for it in the first place? Audiotapes of 160 adult patients' visits to doctors in central North Carolina showed that in only one out of four visits did the patient and physician actually discuss risks. Risk discussion was defined as any discussion about behavior change, tests, or treatments, and their future consequences (such as "You could cut your risk of getting a heart attack in half by . . ."). In the majority of these discussions, the physician stated the risk with certainty (for example, "You will have a heart attack if you don't lose weight"). Only a small proportion of the discussions (about one in six) were initiated by the patient.[16] Moreover, of the 42 patients who said they had discussed risk with their physicians, only 3 could recall the content of the discussion immediately after the visit. But the patients did not seem to mind. More than 90 percent felt that they had had their questions answered, had understood all that was said, and had enough information.

In short, patients felt that their questions had been answered by the doctor, although they asked few questions and remembered even fewer answers. This lack of communication between physicians and patients poses a serious threat to the possibility of "informed" consent.

GEOGRAPHY IS DESTINY

Too often in health care, "geography is destiny." For instance, 8 percent of the children in one community in Vermont had their tonsils removed, but in another, 70 percent had. In Maine, the proportion of women who have had a hysterectomy by the age of 70 varies between communities from less than 20 percent to more than 70 percent. In Iowa, the proportion of men who have undergone prostate surgery by age 85 ranges from 15 percent to more than 60 percent. The *Dartmouth Atlas of Health Care* documents the surprisingly wide variability in the use of surgical treatments across all regions in the United States.[17] Why do these regional differences exist? According to the physician David Eddy, the tendency to follow the local pack is the single most important explanation for regional variations in medical practice.[18] These local customs are fueled by the uncertainty about the outcomes of many surgical treatments. Unlike new drugs, which the Food and Drug Administration assures are tested, surgical procedures and medical devices are not systematically subjected to evaluation.

Aside from geography, the physician's specialization all too frequently determines treatment. The treatment of localized prostate cancer in the United States, for instance, generally depends on whom the patient visits. A study found that some 80 percent of urologists recommend radical surgery, whereas some 90 percent of radiation oncologists recommended radiation treatment.[19] This pattern of variation suggests that patients are not generally advised about their options in a way that encourages them to participate in decision making.

PROSTATE CANCER SCREENING

When the former mayor of New York, Rudolph Giuliani, was diagnosed as having prostate cancer, he was reported in the newspapers to make a pitch that all men participate in prostate cancer screening. "I urge everyone to get the PSA test," Mr. Giuliani said, "If the PSA is normal or low, you don't have a problem. If it's high, then you have."[20] Screening for prostate cancer is typically performed using the prostate-specific antigen (PSA) test or digital rectal examination or both. The case of prostate cancer is a striking example of many men's inability to ask the right kind of questions. For example, I had the following conversation with a friend who is professor of business at a leading American university:

GG: How did you decide whether or not to participate in PSA screening?

Friend: My doctor told me that I am old enough, that it's time, so I went.

GG: Did you ask your doctor about the pros and cons of screening?

Friend: What do you mean? The pros are that the test can detect cancer in an early state.

GG: Did you ask about anything else?

Friend: Not really. The test is so simple, it's cheap, and you can only benefit; it can't hurt.

Note that my friend is an academic who knows how to get information from libraries and the Internet. But rather than using his brain, he displayed respectful docility when he consented to screening. He did not look up information, nor did he ask his doctor the relevant questions: What are the benefits and what are the costs of prostate cancer screening? If he had looked into the relevant medical literature, he would have found that there is *no* evidence that screening reduces mortality from prostate cancer. In other words, people who take PSA tests die equally early and equally often from prostate cancer compared to those who do not. My friend confused early detection with mortality reduction. PSA tests can detect cancer, but because there is as yet no effective treatment, it is not proven that early detection increases life expectancy.[21]

"It can't hurt," responded my friend. He simply assumed that screening has no costs. He erred again; there is no free lunch here. The test produces a substantial number of false positives, and therefore, when there is a suspiciously high PSA level, in most of these cases there is no cancer. That means, many men *without* prostate cancer may go through unnecessary anxieties and often painful follow-up exams. Men *with* prostate cancer are likely to pay more substantial costs. Many of these men undergo surgery or radiation treatment that can result in serious, lifelong harm such as incontinence and impotence. Most prostate cancers are so slow-growing that they might never have been noticed except for the screening. (See Figure 5-2, page 79), which shows that out of 154 people with prostate cancer, only 24 died of the disease.) Autopsies of men older than 50 who die of natural causes indicate that about one in three of them has some form of prostate cancer.[22] More men die *with* prostate cancer than *from* prostate cancer.

Because of the lack of benefit and the likelihood of harm, the U.S. Preventive Services Task Force explicitly does *not* recommend routine screening for prostate cancer, either using prostate-specific antigen (PSA) or digital rectal examinations.[23] There is currently no evidence that early detection reduces mortality, whereas the evidence of possible harm due to follow-up diagnosis and treatment (including incontinence and impotence) is overwhelming. Even when prostate cancer is detected, the prolongation of life by the available treatments is, for all stages of prostate cancer, unproven.[24] Note that in cases of already invasive prostate cancer, treatment can reduce pain; but this should not be confused with reducing mortality. Nevertheless, more and more men are participating in prostate cancer screening, and largely as a result, the number of prostate cancer cases reported in the United States has more than tripled since 1990.

"Doctor's orders!" How, after our conversation, did my friend react? He shook off his respectful docility and looked up the scientific evidence himself. Being an economist by training, he made a calculation of how much money could be reallocated to research effective therapies for prostate cancer if it was not spent on screening. He had taken the first step.

COLORECTAL CANCER SCREENING

To turn the ideal of informed consent into reality, physicians, not only patients, need specific education. Ulrich Hoffrage and I have used natural frequencies to help physicians understand the outcomes of standard tests for colorectal cancer, phenylketonuria, and Bechterew's disease.[25] I will only report the results for the hemoccult test here (also known as FOBT, or fecal occult blood test), which is a standard test for colorectal cancer. The same 48 physicians from Chapter 4—who had an average of 14 years of professional experience—estimated the chance of colorectal cancer given a positive hemoccult test in routine screening. Half of the participants received the information in conditional probabilities, the other half in natural frequencies. The two representations are given here.

Introduction—All Participants

To diagnose colorectal cancer, the hemoccult test—among others—is conducted to detect occult blood in the stool. This test is used from a particular age on, but also in routine screening for early detection of colorectal cancer. Imagine you conduct a screening using the hemoccult test in a certain region. For symptom-free people over 50 years old who participate in screening using the hemoccult test, the following information is available for this region:

Conditional Probabilities Format—First 24 Participants

The probability that one of these people has colorectal cancer is 0.3 percent. If a person has colorectal cancer, the probability is 50 percent that he will have a positive hemoccult test. If a person does not have colorectal cancer, the probability is 3 percent that he will still have a positive hemoccult test. Imagine a person (over age 50, no symptoms) who has a positive hemoccult test in your screening. What is the probability that this person actually has colorectal cancer? _____ percent

Natural Frequencies Format—Remaining 24 Participants

Thirty out of every 10,000 people have colorectal cancer. Of these 30 people with colorectal cancer, 15 will have a positive hemoccult test. Of the remaining 9,970 people without colorectal cancer, 300 will still have a positive hemoccult test. Imagine a sample of people (over age 50, no symptoms) who have positive hemoccult tests in your screening. How many of these people actually have colorectal cancer? _____ out of _____

Figure 6-1 (left side) shows that when the information was represented in probabilities, there was remarkable disagreement among physicians' estimates, which ranged from 1 to 99 percent. The most frequent estimate (50 percent) was 10 times higher than the correct answer, which only 1 out of the 24 physicians reached when they received the information in probabilities. A few others came close to it, but for the wrong reasons. For instance, one physician confused the false positive rate (3 percent) with the probability of colorectal cancer given a positive test, which happened to be not much larger. Thus, we observed the same result as for breast cancer screening: When physicians try to draw a conclusion from probabilities, their minds tend to cloud over.

Did natural frequencies dispel the mental confusion and increase consensus? Again, yes. When the information was expressed in frequencies, the estimates were less scattered, ranging from 1 percent to 10 percent (Figure 6-1, right side). In this group, all of the physicians came up with the correct, or nearly correct, answer. As with breast cancer screening, physicians' clouded thinking about what a positive hemoccult test means can be remedied simply by presenting statistical information differently than it is presented in standard medical textbooks.

The answer to the question "What does a positive test mean?" can be illustrated by drawing a frequency tree (Figure 6-2, page 107). Out of every

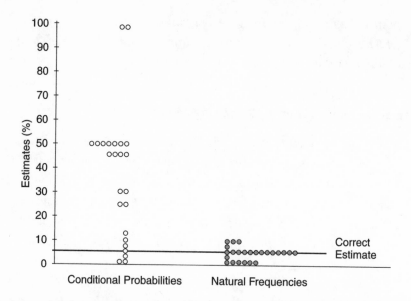

FIGURE 6-1. *The impact of representation on colorectal cancer diagnosis.* Forty-eight physicians estimated the chances of colorectal cancer given a positive screening test. Half of the physicians received the relevant information in conditional probabilities, the other in natural frequencies. Each point represents one physician. The ordinate axis shows physicians' estimates of the probability or frequency of colorectal cancer given a positive test.

315 people who test positive in hemoccult screening, only 15 are expected to have colorectal cancer, which corresponds to a probability of 4.8 percent. Just as in breast cancer screening, most hemoccult tests that are performed for purposes of screening and that turn out positive are false positives. The reason is the same: When a disease is rare, as is colorectal cancer in the general population, the number of true positives will be low and most positives will be false (the exact figures depend on the false positive rate). For instance, one study found that between 94 and 98 percent of patients with positive tests did not have colon cancer.[26] Whether colorectal cancer screening is worth the cost, given this large number of false positives, rests, as with breast cancer screening, in the values of the patient. The physician can assist the patient by explaining clearly what a positive test means.

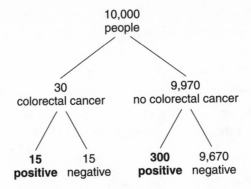

FIGURE 6-2. *Frequency tree for colorectal cancer screening.* Out of every 315 people with a positive hemoccult test (shown in boldface), some 15 will have colorectal cancer. This corresponds to a probability of 4.8 percent. (Data from Gigerenzer, 1996a; Hoffrage and Gigerenzer, 1998.)

Despite the fact that the benefit of colorectal cancer screening seems to be about as high as that for breast cancer, this type of screening has few advocates outside the medical profession. Whereas the vast majority of women who are asked to participate in breast cancer screening comply,[27] according to reports from community screening programs, rates of compliance for the hemoccult test are only 15 to 30 percent, and even lower for sigmoidoscopy, another screening test experienced by patients as uncomfortable, embarrassing, and expensive. If routine hemoccult and sigmoidoscopic screening of all people over the age of 50 were recommended and implemented in the United States, the cost would be more than $1 billion per year in direct charges.

ANATOMY OF THE MEDICAL MIND

How did the physicians go about solving the diagnostic tasks when they gave incorrect answers? From their notes, estimates, and subsequent interviews, we were able to identify most of the intuitive strategies. The reasoning was strikingly different with probabilities than with frequencies. The two dominant strategies among physicians who were given probabilities were "sensitivity only" and "sensitivity minus false positives," both of which are illustrated in Figure 6-3 for the colorectal cancer task. Physicians who reasoned by the sensitivity-only strategy inferred that the chance of colorectal cancer given a positive test was 50 percent, since 50 percent was the sensitivity of the test. Physicians who used the sensitivity-minus-false-positives strategy inferred that the chance was 47 percent, that is, the sensi-

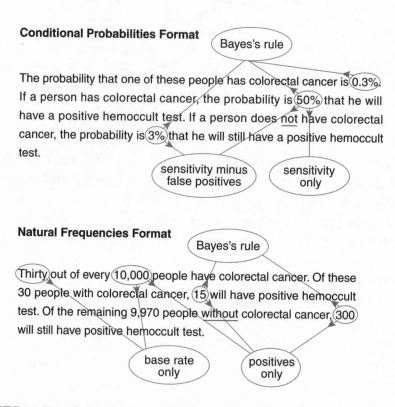

Conditional Probabilities Format

Bayes's rule

The probability that one of these people has colorectal cancer is 0.3%. If a person has colorectal cancer, the probability is 50% that he will have a positive hemoccult test. If a person does not have colorectal cancer, the probability is 3% that he will still have a positive hemoccult test.

sensitivity minus false positives

sensitivity only

Natural Frequencies Format

Bayes's rule

Thirty out of every 10,000 people have colorectal cancer. Of these 30 people with colorectal cancer, 15 will have positive hemoccult test. Of the remaining 9,970 people without colorectal cancer, 300 will still have positive hemoccult test.

base rate only

positives only

FIGURE 6-3. *How physicians reason.* The arrows show which information is used by the most common diagnostic strategies. For instance, physicians who use the strategy "sensitivity only" infer that the probability that the patient has colorectal cancer given a positive test is 50 percent, which is the sensitivity of the test. Note that physicians' reasoning strategies change when the information is presented in natural frequencies (bottom) rather than conditional probabilities (top). The analysis is based on 48 physicians. (See Hoffrage and Gigerenzer, 1998.)

tivity of the test minus the false positive rate (3 percent).[28] In general, these two strategies lead to gross overestimates of the actual risk of someone's having a disease given a positive test result—because both strategies ignore the disease base rate (0.3 percent) and most diseases are relatively rare.

How consistently did physicians apply these strategies—for instance, did a physician who used the sensitivity-only strategy to solve one diagnostic task also use it to solve the others? When confronted with probabilities, the physicians showed little consistency. For instance, one department chief added the sensitivity to the false positive rate to infer the chance of cancer given a positive test, and in the next diagnostic task he multiplied the sensitivity by the disease base rate. Only 1 out of 5 physicians used the same strategy to solve both probability tasks that they received, such as the breast cancer and colorectal cancer tasks. With frequencies, consistency increased: More than half of the physicians used the same strategy in both tasks.[29]

Although the physicians were much more likely to answer correctly in the tasks expressed in frequencies, even there their estimates were sometimes incorrect. With frequencies, the two most frequent non-Bayesian strategies were "base rate only" and "positives only." In the colorectal cancer task, the base-rate-only strategy leads to the inference that 30 out of 10,000 people with a positive hemoccult test actually have colorectal cancer, and the positives-only strategy leads to an estimate of 315 out of 10,000. The first is an underestimation of the actual risk; the second (about 3 percent) is close to the right estimate, but for the wrong reason. The two strategies follow a common logic, however: Both focus on one of the two base rates—either that of the underlying disease or that of the observable result (a positive hemoccult test). Thus, the use of natural frequencies not only fosters correct Bayesian inferences but, barring that, encourages the use of strategies that rely on base rates.

On the basis of physicians' poor diagnostic reasoning using probabilities, previous researchers concluded that physicians confuse the two conditional probabilities, namely, the test sensitivity and the probability of the disease given a positive test.[30] The analysis of the medical mind that Ulrich Hoffrage and I made provides some empirical evidence for this hypothesis (the sensitivity-only strategy), but it also reveals that multiple inappropriate strategies—not just this one—are sometimes used in diagnostic inference. When information is presented in probabilities, the high variability in diagnostic judgments can be traced to the high variability in diagnostic strategies.

We also found in our studies that the younger physicians are better at thinking statistically than their older counterparts. Dr. Standing—the department chief we met in Chapter 4—was not the only one who, despairing of his ability to make a diagnostic inference using statistical information, referred to a daughter or a son as someone who would know better. During one of the interviews conducted for our study, the 18-year-old daughter of a 49-year-old private practitioner happened to come in and asked to take the test too. Her father had worked for 30 minutes and failed on all four diagnostic tasks, shifting desperately between two strategies (sensitivity-only and base-rate-only). In his case, even natural frequencies were no help. His daughter, in contrast, solved all four tasks by drawing frequency trees such as those shown in Figures 4-2 and 6-2. When she discovered which strategies her father had been using, she gave her father a quizzical look and said: "Daddy, look, the frequency problem is not hard. You couldn't do this either?"

Are physicians aware of their innumeracy? In general, yes, but they tend not to be aware that proper representations can turn their innumeracy into insight. A few physicians even celebrate their ignorance, in an attitude reminiscent of the physician-as-artist. For instance, a university professor who was an ear, nose, and throat specialist—one of three physicians who declined to participate in our study—declared, "This is not the way to treat patients. I throw all these journals [with statistical information] away immediately. One can't make a diagnosis on such a basis. Statistical information is one big lie." However, the majority of physicians in our studies who declared themselves to be mathematically illiterate had feelings of regret: "But this is mathematics. I can't do that. I'm too dumb for this." "I can't do much with numbers. I am an intuitive being. I treat my patients in a holistic manner and don't use statistics." With natural frequencies, to their own surprise, these self-diagnosed innumerate physicians reasoned as well as those of their fellows who did not claim to suffer from innumeracy. Their nervousness and tension turned into relief. "Now it's different. It's quite easy to imagine. There is a frequency; that's more visual," said one, and another said, "A first grader could do this. Wow, if someone couldn't solve this!"

With reference to the desirability of informed consent, it must be realized that such consent is about more than just signing a form; it is about risk communication. This fact should have consequences for medical training.[31] Every medical student needs to be trained in the use of mind tools that facilitate communication. And judging from the feedback we got from the physicians in our studies, such training would not be unwelcome. After participating, one of them wrote to us: "Participating in this study and learning its results is of great importance to me professionally. I'm sure that from now on I will represent medical data to myself in frequencies rather than just glancing over them or being content with some vague idea."

Why Informed Consent Is Not Easy to Achieve

Once I had such a stiff neck that I could hardly turn my head. My chiropractor sent me to a radiologist, Dr. Best,[32] for an X ray. Dr. Best, a gentle man with empathetic eyes, ran a large radiology institute that also offered mammography screening. White-clad assistants whirled through the hallways, and the waiting rooms overflowed with patients. After my X ray was taken, Dr. Best mounted it on a lighted screen and explained to me what he saw. I soon realized that my chiropractor had told him on the phone that I study decision making under uncertainty. Dr. Best was overjoyed to have someone to talk to about his work: "You cannot imagine how boring my life is, the same routine every day—X rays for 25 years." He complained about patients' fears of X rays, about having to perform all kinds of analyses on patients to protect himself from being sued for malpractice, and about many patients' inclination to delegate every bit of decision making to the doctor.

As we discussed fear and responsibility, I took the opportunity to ask him what he would do in the following situation: "Imagine I were a woman in her early 40s who came to your institute for breast cancer screening—not because I had any symptoms, just because my regular doctor told me to do so every other year. Now suppose that the mammogram came out positive, and I wanted to know what the chances are that I actually have breast cancer. What would you tell me?" "That it looks like you have breast cancer, but it's not certain," answered Dr. Best, who went on to emphasize the importance of empathizing with patients. His credo was never to deny the patient hope. I reminded him of the actual probabilities involved (see Chapter 4), which he knew, and explained that it follows that only about 1 out of 10 asymptomatic women who test positive actually have breast cancer. He looked at me and said: "Hmm. . . . You know, at my university, we never learned how to think with probabilities. And now, just look in the waiting room—I don't have time to read professional journals after a 12-hour workday." And on he went to explain how to distinguish between patients who can bear the report of a positive test and those with whom he has to be more sensitive about disclosing the diagnosis.

After 15 minutes of talking about patients' personalities and physicians' empathy, I posed the question about breast cancer screening to Dr. Best again. "What would you tell a woman in this situation?" I asked. "The truth," he answered. "What is the truth?" I rejoined. "That it looks like she has breast cancer." "But 15 minutes ago," I reminded him, "we just agreed that only about 1 in 10 asymptomatic women who test positive actually has breast cancer, that is, an asymptomatic woman who tests positive is more likely not to have breast cancer than to have it." "That is also true," Dr. Best replied. "We physicians should be taught about these things. But when can I take the time? It's all a cost-benefit question."

The reader may ask how one could perform mammography screening for 25 years and not notice that most women who test positive do not have breast cancer. Similarly, the majority of physicians in Figure 4-1 on page 43 (left side) also did not seem to have noticed. If physicians and patients do not know the facts, there may be consent, but not informed consent. The possibility of informed consent does not—or does not only—depend on the patient's intelligence, maturity, and ability to cope; it also depends on the constraints under which physicians work. The 12-hour workday of Dr. Best is such a constraint. What are the major constraints that prevent physicians such as Dr. Best from realizing that most women who test positive do not have cancer? More generally, what are the institutional constraints that work against the ideal of informed consent?[33]

Division of labor. First, there is a division of labor, which can obstruct the flow of information. Radiologists who perform mammography examinations typically do not find out whether a patient later develops cancer. Most health care systems do not monitor and disseminate follow-up information, and there is little incentive for physicians to try tracking down the numbers themselves. This explanation applies to radiologists, such as Dr. Best, but not, however, to gynecologists, who do see the relevant information.

Legal and financial incentive structure. The second reason concerns professional fear and pride, and the legal and financial incentives associated with it. The error physicians fear most is to miss a cancer—the emotional distress at having had the power to detect a cancer and missing the opportunity. A miss can damage their reputation; fellow physicians may take notice. Equally important, a miss also makes them vulnerable to being sued. Erring on the side of overestimating the chances of cancer protects physicians from being sued because it means they will rarely miss one. At the same time, this policy brings in more revenue to hospitals and private practices, owing to additional diagnoses and treatment. The costs of this policy—a large number of false positives and their potential physical, psychological, and monetary costs to the patient —vanish in the face of the physician's fear of missing a cancer. As the organizer of the earlier discussion mentioned, female patients are generally grateful for false positives. However, women might be less grateful for what turn out to be false positives (and less terrified to begin with) if they were informed that some 9 out of 10 women who test positive in screening do not have breast cancer in the first place.

Conflicts of interest. The third reason is various conflicts of interest. A breast cancer specialist told me that he no longer routinely recommends that women visit a radiologist just "because it is time and every woman should go." Instead, he decided to inform each woman about the benefits and costs of mammography screening so that she can make a reasoned decision as to whether and when to participate. When he spoke to a radiologist friend over dinner about his change in policy, the friend got so upset that he dropped his knife and fork, stood up to leave the restaurant, and exclaimed, "Where did you get these numbers?" "From several hundred thousand women studied in America, Sweden, and other countries," was the reply. "In America," the radiologist exclaimed in anger, "they don't know how to read mammograms!" But his problem in reality was not with America; it was economic. For years, he had screened the women whom breast cancer specialists had sent to him. If half of the women were to decide not to participate in screening or to begin participating at a later age, the radiologist would face financial ruin. I admire the breast cancer expert for his willingness to risk losing outraged friends for the sake of informing his patients.

Innumeracy. Last but not least, there is innumeracy. Many physicians are poorly educated in statistical thinking and have little incentive to engage in this alien form of reasoning. If patients start to look up numbers, physicians might be forced to do so too.

The first three reasons for many physicians' failure to inform patients arise from institutional, professional, and economic structures that are beyond the power of this book to change. But as for the fourth reason—innumeracy—there is cause for hope: This book presents highly effective and inexpensive, simple tools for turning innumeracy into insight. Once a sufficient number of physicians and patients master these tools, the insight they gain will put pressure on the institutional, professional, and economic structures to change.

A positive result means antibodies to HIV were found in your blood. This means you have HIV infection. You are infected for life and can spread HIV to others.

Illinois Department of Public Health

I will kill myself if I test positive.

A client

7

AIDS COUNSELING

Betty

One day in November 1990, Betty's phone rang. She lived in Florida and was 45 years old and the mother of three teenage sons; their father had died. She was asked to come to the local health clinic, where she had had a checkup for a thyroid problem and a blood sample had been taken for testing. When she arrived, she was told she had AIDS. The doctors were not sure how long she had to live. In the months that followed, she watched television constantly to block out thoughts of the disease. But the thoughts came back during the night: What dress do I want to be buried in? How are my kids going to take it? How will people treat them?

In 1992, her doctor put her on didanosine (an anti-HIV replication drug), which caused vomiting, fatigue, and other side effects. When she joined a local group for AIDS patients, the counselors noted that her T-cell count had remained consistently high. They suggested that she be retested.

In November 1992, Betty's phone rang and she was again asked to come to the clinic. When she arrived, she was told, "Guess what? Your HIV test came out negative!"

Betty sued her doctor, the clinic, and the Florida Department of Health and Rehabilitation Services, the agency that had performed the initial test. A jury awarded her $600,000 for two years of pain and suffering.[1]

David

The *Chicago Tribune* published the following letter and response on March 5, 1993, under the headline "A False HIV Test Caused 18 Months of Hell":

Dear Ann Landers,

In March 1991, I went to an anonymous testing center for a routine HIV test. In two weeks, the results came back positive. I was devastated. I was 20 years old and doomed. I became severely depressed and contemplated a variety of ways to commit suicide. After encouragement from family and friends, I decided to fight back. My doctors in Dallas told me that California had the best care for HIV patients, so I packed everything and headed west. It took three months to find a doctor I trusted. Before this physician would treat me, he insisted on running more tests. Imagine my shock when the new results came back negative. The doctor tested me again, and the results were clearly negative.

I'm grateful to be healthy, but the 18 months I thought I had the virus changed my life forever. I'm begging doctors to be more careful. I also want to tell your readers to be sure and get a second opinion. I will continue to be tested for HIV every six months, but I am no longer terrified.

David in Dallas

Dear David:

Yours is truly a nightmare with a happy ending, but don't blame the doctor. It's the lab that needs to shape up. The moral of your story is this: *Get a second opinion. And a third.* Never trust a single test. Ever.

Ann Landers

David does not say what, if anything, his doctors told him about the chances that he actually had the virus given the positive HIV (human immunodeficiency virus) test result. He seems to have inferred that a positive test meant that he had the virus, period. Betty was simply told that she had AIDS. Susan, the single mother we met in Chapter 1, was told that a positive test meant, with absolute certainty, that she had HIV. Susan had unprotected sex with an HIV-infected person, believing that it did not matter because she already had the virus; Betty's life turned into two years of suffering; David contemplated suicide—only to learn the hard way that HIV tests can result in false positives.

What does a positive HIV test mean? A negative result? And how can a counselor communicate this to a client, so that he or she understands the result? In this chapter, I address these questions for the testing of people who do not practice risky behavior, such as the use of IV drugs. First, however, let us have a closer look at the test, the disease, and the social stigma attached to HIV positives.

HIV and AIDS

When is a test result declared positive? HIV testing typically involves the following sequence. The first test, called the ELISA (enzyme-linked immunoabsorbent assay), is designed to detect antibodies against HIV in blood samples. It was originally used to screen donated blood, where maximizing test sensitivity (the true positive rate)—at the cost of an increased false positive rate—was imperative. If the result is negative, then the client is notified that he or she is HIV-negative. If the result is positive, then at least one more ELISA (preferably from a different manufacturer) is conducted on the sample. If the result is still positive, a Western blot test, which is more expensive and time-consuming to conduct than ELISA, is performed. If the Western blot is also positive, the client is notified that he or she is HIV-positive. In some cases, a second blood sample is obtained and analyzed before the client is notified.[2] The exact procedure varies across institutions and countries.

AIDS (acquired immune deficiency syndrome) is defined primarily by a severe immune deficiency. Unlike other diseases, it has no constant, specific symptoms. Once the immune system has begun to malfunction, a broad spectrum of health complications can set in, and there are some 26 opportunistic infections known. If a person tests positive for HIV *and* has one or more of these infections, she or he is diagnosed with AIDS. AIDS is the final stage of a viral infection caused by HIV (but individuals can express AIDS for reasons other than HIV). HIV is a retrovirus—that is, a virus that inserts its genetic material into that of the human host cell, probably for the lifetime of the host. It destroys the T cells in the immune system. Two strains are distinguished, HIV-1, discovered in 1983 and the cause of most AIDS cases worldwide, and HIV-2, discovered in 1987 in West African women. HIV-2 is rare in the United States and in Europe. HIV-2 seems to be less harmful to the immune system and reproduces more slowly.[3]

Is there a cure? Not at present. Some of the problems associated with finding a cure can be illustrated by a comparison with the syphilis epidemic in the early part of the twentieth century. The syphilis campaigns closely paralleled today's AIDS campaigns. There were educational programs to reduce high-risk sexual behavior, scare tactics were spread through the media, and serological testing was made mandatory before one could obtain a marriage license in certain states in the United States. All of these measures, however, had little effect on the spread of the epidemic. By the 1930s, almost 1 in 10 Americans was infected with syphilis. The syphilis epidemic was finally brought under control—not by changes in human sexual behavior, but by the discovery of penicillin, a cheap and effective drug.

The important difference between syphilis and AIDS is that the bacterium that causes syphilis (a spirochete) does not mutate and change as rapidly as HIV does. When HIV replicates itself, it makes so many copying errors that by the time a person is diagnosed with AIDS, that person may have a billion or more HIV variants in his or her body. Some of these mutations will weaken HIV and expose it to attack by the immune system; others will strengthen HIV, increasing its chances of evading the immune system. This rapid Darwinian-type evolution of the virus seems to exceed the capacity of the immune system to recognize and respond to it, and it helps the virus become resistant to drug therapies. Between the time of the infection and the illness itself, there is an asymptomatic phase of ten to twelve years on average; however, this may not be a real latency phase, but a time of continuous struggle between HIV and the immune system, with the balance slowly shifting in favor of the virus.[4]

There is no cure, but there is hope. Drugs that interfere with the virus's ability to replicate have been developed. Because the virus can rapidly become resistant to each of them, a combination of drugs is used. This so-called AIDS drug cocktail therapy can prolong the life of an infected person, but it is not a cure. It uses a mixture of drugs, including didanosine and zidovudine. The downside of this positive development is that these cocktails are very expensive and are therefore more readily available to the rich than to the many in poverty; in addition, these drugs have severe to moderate side effects, including burning pains in the hands and feet, hair loss, and a dangerous swelling of the pancreas.

Screening for HIV is most important for high-risk groups. Although the possibility of therapy is very limited because of the relentless course of the disease, there is another reason for screening. Early detection can reduce the prevalence of HIV. Recall that mammography does not reduce prevalence, only mortality. HIV, however, is different because it is transmitted from person to person. Early detection can reduce the spread of the virus, and thereby its prevalence, if the infected persons disclose their status to their sexual partners and take precautions.

Stigma and Sexual Ethics

Ryan White was 12 years old when he was diagnosed with AIDS. He was a hemophiliac, and he lived in Kokomo, Indiana, where the disease was socially unacceptable. He got the virus from a blood transfusion essential to his survival. Ryan suffered from the acrimony and lies of his classmates and their parents, who accused him of spitting on them to infect them with the virus, among other, similar fabrications. Ryan said he understood that this discrimination was due to ignorance, fear, and misconceptions about how the virus is transmitted. His fight to be allowed to attend school, to be able to leave his home and walk about without being ridiculed, and to become socially acceptable gained him the respect of millions across the United States. Ryan died in 1990 at the age of 18.

The first AIDS cases in America (in 1981) were gay men. The Reverend Billy Graham said, "AIDS is a judgment of God."[5] Initial public reaction was denial: by the media, which were not ready to talk about homosexuality, needles, or condoms, as well as by gay men, who were not ready to rethink their sexual liberation won in the 1970s. A 1988 *Cosmopolitan* magazine article by Robert Gould assured women that there was practically no risk of AIDS with ordinary vaginal or oral sex—even when the partner was HIV-positive.

The event that stopped the denial happened in 1985. The Hollywood star Rock Hudson disclosed in public that he was suffering from AIDS. Other celebrities followed. The pianist and entertainer Liberace died two years later of AIDS, and the professional basketball player Magic Johnson announced in 1991, "Here I am saying it can happen to anybody, even me, Magic Johnson."

Today, the virus is spread worldwide primarily through heterosexual contact. However, this insight has not halted the stigmatization of the victims. In Hinton, West Virginia, one woman was killed by three bullets and her body dumped beside a remote road; another woman was beaten to death, run over by a car, and left in the gutter. Both had told people that they had AIDS, and the authorities said that this was the reason they had been killed. In Ohio, a man with a positive HIV test lost, within 12 days, his job, his home, and—almost—his wife. The day he was going to commit suicide he was notified that he had received a false positive test result.

Others choose not to speak up, but rather to deceive or to lie. Many HIV-infected individuals do not disclose their status to their sexual partners. For instance, patients who were in primary care for HIV infection in Boston City Hospital and Rhode Island Hospital were asked whether they had disclosed their infection to their sexual partners. Forty percent said that they did not tell their partners, and most of them did not use condoms all the time.[6] Women disclosed more often than men did, and individuals with one sexual partner more often than those with multiple partners. Southern California college students report similar deceptive behaviors. Among some 500 sexually experienced students, 47 percent of the men and 60 percent of the women reported that they had been told a lie in order to be induced to have sex. However, only 35 percent of the men and 10 percent of the women said that they had lied in order to have sex. That is, more report having been the victim rather than the perpetrator—which is only consistent with the extent of lying and deception admitted. One in five male students said that if he were HIV-positive, he would lie and claim to have a negative HIV test in order to have sex.[7] The virus can take advantage not only of human lies to spread itself, but also of myths. At the 13th International AIDS conference in July 2000, Dr. Zweli Mkhize, the health minister of a South African province, urged for education to combat false but widespread beliefs, such as that having sex with a virgin will cure AIDS, a myth that has led to violence and rape.[8] This is a contemporary version of the ancient myth of the power of virgins: only someone so pure can absorb pollution.

The Illusion of Certainty

At a conference on AIDS held in 1987, former Senator Lawton Chiles of Florida reported that of 22 blood donors in Florida who were notified that they had tested HIV-positive with the ELISA, 7 had committed suicide. (At that time, ELISA plus Western blot was not yet standard procedure.) The medical text documenting this tragedy many years later informed the reader that "even if the results of both AIDS tests, the ELISA and WB [Western blot] are positive, the chances are only 50-50 that the individual is infected."[9] This figure applies when people in a low-risk group—such as blood donors who are screened and selected for not having infectious diseases—test positive. The unfortunate donors mentioned by Senator Chiles were tested only with the ELISA, which has a higher false positive rate than the combination of the ELISA and the Western blot. Thus, the chance that they were infected was probably even lower than the 50 percent estimated by the text. If the donors had been informed of their actual risk of having HIV given a positive test, some or all of the seven unfortunate donors might still be alive today.

Not knowing about the possibility of false positives is one form of the illusion of certainty, but it is not the only one. The feeling "it cannot happen to me" is another. According to several studies, half of American teenagers are not worried about being infected with HIV and consequently do not change their sexual behavior. At the same time, one out of four new HIV infections occurs in people between the ages of 13 and 20. As one teenager explained, "When you're a teenager, your hormones are raging and you think you're indestructible. But sex is how I got AIDS."[10] Two groups, adolescent gay men and teenage women infected via heterosexual contact, make up some 75 percent of all teenage HIV infections. Both forms of the illusion of certainty—not realizing that tests can err and not realizing that one can be vulnerable—have similar consequences: some people have contemplated suicide, some have committed suicide, and others have turned to a fatalistic or careless lifestyle that endangered themselves and others.

We may not be able at present to win the war against the virus with biological weapons, but we can help people to understand the risks better. Better understanding can reduce some of the toll the disease takes every year. As the cases of Susan, Betty, David, and the Florida blood donors indicate, there seems to be a problem with counseling. I invite you to take a closer look into the counseling room, at the counseling of people who do not practice high-risk behavior.

Low-Risk Clients

Populations with the highest risk of HIV infection are homosexual men, intravenous drug users, heterosexual partners of intravenous drug users, hemophiliacs, and children of HIV-infected mothers. However, a large proportion of people at high risk choose not to be tested, whereas those at low risk increasingly undergo HIV tests. In the United States alone, about 50 million blood and plasma samples are tested each year.[11] About 60 percent of the general population in Switzerland has had at least one HIV test.[12] A large number of these are people engaging in low-risk behavior.

People at low HIV risk have HIV tests for various reasons: voluntarily, because they want to find out whether they are infected before starting a relationship, getting married, having children, or for other reasons; and involuntarily, because they are blood donors, immigrants, applicants for health insurance, military personnel, or members of other groups required by law to take the test. For instance, a friend of mine and her fiancé chose to be tested for HIV before they got married—just to be sure. The Swedish government has encouraged voluntary testing to the point that "people who are unlikely to be infected are the ones who take the test, in droves."[13]

Involuntary testing is a legal possibility in several countries, and insurers can exploit this to protect themselves against losses. For instance, in 1990, Bill Clinton, then governor of Arkansas, had to take an HIV test to get his life insurance renewed. In the late 1980s, Illinois and Louisiana required couples applying for marriage licenses to be tested for HIV, and disclosed the results to both partners. These programs incurred great social and financial cost; they falsely identified some people as being infected with HIV, resulting in broken engagements, aborted pregnancies, and psychological distress.[14] The American Medical Association has endorsed mandatory HIV testing of all pregnant women and newborn babies, and New York, in 1996, was the first state to pass legislation mandating HIV testing of newborns—known as the "Baby AIDS" bill.[15] Compulsory testing has been forced by courts on individuals, and by governments on prisoners, prostitutes, and persons seeking to immigrate. People at low risk may even be subjected to HIV tests without their knowledge—for instance, large companies in Bombay reportedly tested their employees for AIDS without telling them; when the tests came out positive, the employees were fired.

Counseling people at low HIV risk requires giving particular attention to false positives, that is, to the possibility that the client may receive a positive HIV test result despite not being infected. The reason is that the lower the prevalence (base rate) of HIV in a given group, the larger the proportion of false positives among the positive results. In other words, if someone with high-risk behavior tests positive, the probability that he is actually infected with HIV is very high, but if someone in a low-risk group tests positive, this probability is considerably lower.

What Does a Positive Test Mean?

A number of years ago, as a citizen of Germany, I applied for a U.S. green card so that I could accept a professorship at the University of Chicago. The U.S. immigration office demanded that I have blood taken for an HIV test and informed me that a positive result would mean a denial of the green card. So one morning I drove to the U.S. consulate in Frankfurt to have the test done. On the way, I asked myself, How likely is it that a man who tests positive for HIV (ELISA and Western blot test, one blood sample) actually has the virus? At this time, I had the following information concerning German men with no known risk behavior:

> *About 0.01 percent of men with no known risk behavior are infected with HIV (base rate). If such a man has the virus, there is a 99.9 percent chance that the test result will be positive (sensitivity). If a man is not infected, there is a 99.99 percent chance that the test result will be negative (specificity).*

What is the chance that a man who tests positive actually has the virus? Most people think it is 99 percent or higher—however, their minds are clouded by the probabilities. One way to find the answer is to take pencil and paper and insert these probabilities into Bayes's rule—but I was driving. Even on the autobahn, however, it is easy to mentally transform this information into natural frequencies:

> *Imagine 10,000 men who are not in any known risk category. One is infected (base rate) and will test positive with practical certainty (sensitivity). Of the 9,999 men who are not infected, another one will also test positive (false positive rate). So we can expect that two men will test positive.*

How many of the men who test positive are actually infected? Using this mental representation, I could easily see that my chances of having the virus—given a positive test result—would be approximately 1 in 2, or 50 percent. The tree in Figure 7-1 illustrates this result. That is, testing positive for HIV would be no reason to contemplate suicide or a move to California, nor should such a result by itself be sufficient to bar entry to the United States. It would be a reason to have another test with a new blood sample.

How could Susan, Betty, and David have been spared their nightmares? How could the suicides in Florida have been prevented? The answer is, by

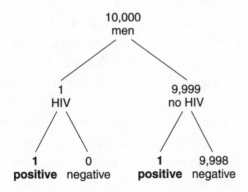

FIGURE 7-1. *What does a positive HIV test mean?* Out of 10,000 men with no known risk behavior, two will test positive (shown in boldface) and one of these will have the virus. (Data from Gigerenzer et al., 1998)

counselors using the same transparent risk communication as I did while driving. If a client had practiced risky behavior—for example, if David were homosexual and belonged to a group with a 1.5 percent base rate of HIV infection—transparency could be achieved in the same way. In that case, the physicians might have explained:

> *Think of 10,000 homosexual men. We expect 150 to be infected with the virus, and most likely all of them will test positive. Of the 9,850 men who are not infected, we expect that 1 will test positive. Thus, we have 151 men who test positive, of whom 150 have the virus. Your chances of not having the virus are therefore 1 out of 151, that is, less than 1 percent.*

This would be bad news—only slightly better than leaving the counselor's office with the belief that it is absolutely certain that one has an HIV infection. One can see that the meaning of a positive test result depends on the *reference class,* which determines the base rate of HIV. If the reference class included only gay men who practice safe sex, the chances of not having the virus after a positive test result would be better.

What about Ann Landers's comment? Her advice to "get a second opinion" is right to the point. But her response that David should not blame the doctor but rather the lab overlooks the fact that whatever the reasons for false positives, doctors should inform patients that false positives occur and should say about how often. Some, but not all, false positives can be traced to laboratory errors, which include blood samples being confused or contaminated in the lab and errors when entering the data into the computer (as in Susan's case in Chapter 1). False positives can also result from certain medical conditions that have nothing to do with HIV, such as rheumatological diseases, liver disease, various cancers, malaria, and alcoholic hepatitis.[16] The estimates of a sensitivity of 99.9 percent and a specificity of 99.99 percent are the best I know of for combined ELISAs and Western blot (one blood sample), but they are approximations.[17]

Is the inadequate counseling that Susan, Betty, and David seem to have received an exception, or is it the rule? How do professional AIDS counselors communicate risks to their clients?

Inside the Counseling Room

I have been lucky in having not only smart students, but also brave ones. To find out firsthand how risk is communicated, a student of mine, Axel Ebert, volunteered to go undercover to 20 public health centers to have 20 HIV tests.[18] The centers were located in 20 German cities, including the three largest German cities of Berlin, Hamburg, and Munich, and they offer free HIV tests and counseling to the general public. Pretest counseling is mandatory, and this allowed Ebert to ask the relevant questions, such as "Could I ever test positive if I do not have the virus? And if so, how often does this happen?"

Ebert first contacted the health centers by phone and made an appointment. He was able to visit two centers in succession. Then he waited at least 2 weeks to allow the bruises from the perforation of his veins in both arms to heal. These pauses were necessary, because needle marks would have suggested to a counselor that Ebert was a drug addict and therefore in the high-risk category.

Of the 20 professional counselors, 14 were physicians and the others were social workers. Counseling before HIV testing is intended to help the client understand the testing procedure, the risks of HIV infection, and the meanings of a positive and a negative result. A report of the German government explicitly guides counselors to perform a "quantitative and qualitative assessment of the individual risk" and to "explain the reliability of the test result" before the test is performed.[19]

Ebert asked each counselor the following questions, unless the counselor offered the information spontaneously:

> *Sensitivity.* If one is infected with HIV, is it possible to receive a negative test result? How reliably does the test identify the virus if it is present?
> *False positives.* If one is not infected with HIV, is it possible to receive a positive test result? How reliable is the test with respect to false positive results?
> *Prevalence in low-risk clients.* How frequent is the virus in my risk group, that is, among heterosexual men between 20 and 30 years old with no known risk factor such as intravenous drug use?

Positive predictive value. What are the chances that men in my risk group actually have HIV given a positive result?

The *positive predictive value* is the probability of being infected with HIV given a positive test. During the counseling session, Ebert never used technical terms such as "positive predictive value" but everyday language, as illustrated above. When the counselor gave a quantitative answer (a number or a range) or said that he or she could not give a more precise answer, then Ebert went on to another question. If the answer was qualitative (for example, "fairly certain") or if the counselor misunderstood or avoided answering the question (for example, "don't worry, the test is very reliable, trust me"), then Ebert asked for further clarification and, if necessary, repeated the request for clarification one more time. If the third attempt failed, he did not push further, because some physicians get defensive or angry when clients continue to insist on clarification. When Ebert asked about the prevalence of HIV and about the positive predictive value, he always reminded the counselor of his low-risk status (a 25-year-old heterosexual man who does not use intravenous drugs and has no other known risk factors).

Ebert used a coding system to record the relevant information in shorthand during each counseling session. After each session but three, he had an HIV test performed. In two cases he would have had to wait several hours to have the test, and in one case the counselor suggested that he sleep on it before deciding whether to be tested. Investigating AIDS counselors' behavior without telling them that they are being studied raised ethical questions; we, therefore, obtained the clearance of the Ethics Committee of the German Association of Psychology. We apologize to the counselors for having employed this covert method, but believe that the results of this study justify it. They indicate how AIDS counseling can be improved.

COUNSELING SESSIONS

Let us first look at a few individual counseling sessions and then at the overall results. Session 1 took place in 1994 in a public health center in a city with a population of about 200,000. Ebert's questions are in italics and abbreviated using the technical terms. The counselor's answers are marked in the subsequent text. If there is more than one answer to a question, these are answers to Ebert's clarifying questions.

Session 1: The Counselor Was a Female Social Worker

Sensitivity?

- False negatives really never occur. Although, if I think about the literature, there were reports of such cases.

- I don't know exactly how many.

- It happened only once or twice.

False positives?

- No, because the test is repeated; it is absolutely certain.

- If there are antibodies, the test identifies them unambiguously and with absolute certainty.

- No, it is absolutely impossible that there are false positives; because it is repeated, the test is absolutely certain.

Prevalence?

- I can't tell you this exactly.

- Between about 1 in 500 and 1 in 1,000.

Positive predictive value?

- As I have now told you repeatedly, the test is absolutely certain.

The counselor was aware that HIV tests can lead to a few false negatives, but incorrectly informed Ebert that there are no false positives. Ebert asked for clarification twice, in order to make sure that he correctly understood that a false positive is impossible. The counselor asserted that a positive test result means, with absolute certainty, that the client has the virus; this conclusion follows logically from her (incorrect) assertion that false positives cannot occur. In this counseling session, Ebert was told exactly what Susan had been told by her Virginia physicians: If you test positive, it is absolutely certain that you have the virus. Period.

The next session took place in a city with a population of about 300,000.

Session 2: Counselor Was a Male Physician

Sensitivity?

- When there are enough antibodies, then the test identifies them in every case. Two tests [ELISA and Western blot] are performed; the first test is in its fourth generation and is tuned to be very specific and sensitive. Nevertheless, it is tuned in such a way that it is more likely to identify positives than negatives.

- 99.8 percent sensitivity and specificity. But we repeat the test, and when it comes out positive, then the result is as solid as cast iron.

False positives?

- With certainty, they do not occur; if there are false results, then only false negatives, occurring when the antibodies have not yet formed.

- If you take the test here, including a confirmatory test, it is extremely certain: in any case the specificity is 99.7 percent. This is as solid as cast iron. We eliminate confusion by using two tests.

Prevalence?

- The classification of individuals into risk groups is now outdated; therefore one cannot look at this that way.

- I don't remember this. There is a trend in which the virus is spreading into the general public. Statistics are of no use for the individual case!

Positive predictive value?

- As I already have said: extremely certain, 99.8 percent.

This counselor initially denied the existence of false positives. Unlike the counselor in Session 1, however, he changed his mind when the client asked for clarification and estimated the false positive rate for the combination of ELISAs and Western blot to be 0.3 percent (this follows from a specificity of 99.7 percent). This rate is much higher than the literature indicates. When the counselor estimated the positive predictive value, he confused it with the sensitivity, as reflected by the phrase "as I already have said." As a consequence, the information he gave is contradictory: it is internally inconsistent. We can see through this confusion by translating this counselor's probabilities into natural frequencies, using as the base rate (which he did not specify) the median estimate of all counselors, 1 in 1,000. Think of 1,000 low-risk male clients. One has the virus and will test positive with practical certainty. Of the remaining 999 uninfected men, 3 will test positive. Thus we expect that out of 4 men who test positive, only 1 has HIV. One in four is not 99.8 percent.

Unlike the first counselor, who suffered from an illusion of certainty, the second counselor did not deny the existence of false positives. However, he did not know how to express the risks so that he and his clients could understand them. His way of talking about risk was to use conditional probabilities, expressed as percentages, and his mind was clouded. This counselor did not even realize that his figures were impossible.

The phrase "when the antibodies have not yet formed" refers to the *window period,* that is, the time interval between infection and the formation of a sufficient number of antibodies to be detected by the tests. For instance, infection by sexual intercourse has an average window period of about 6 months.[20] During this period, false negatives are likely.

Session 3 took place in a city with more than 1 million inhabitants.

Session 3: The Counselor Was a Female Physician

Sensitivity?

• The test is very, very reliable, that is, about 99.98 percent.

False positives?

• The test will be repeated. After the first test, one does not speak of positive, only of reactive. When all tests are performed, then the result is certain.

• It is hard to say how many false positives occur.

• How many precisely? I would have to look in the literature to see if I could find this information there.

Prevalence?

• That depends on the region.

• Of the circa 67,000 infected people in Germany, 9 percent are heterosexual.

• In this city we have 10,000 infected people, that is, 1 percent of the population. But these are only numbers that tell you nothing about whether you have the virus or not.

Positive predictive value?

• As I already mentioned, the result is 99.98 percent sure. If you receive a positive result, you can trust it.

As in the previous session, the counselor at first suggested that there are no false positives. When Ebert asked for clarification, however, she made it clear that false positives exist, but that she could not say how many. Like the counselor in Session 2, she confused the sensitivity, that is, the chance that an infected person receives a positive result, with the positive predictive value, that is, the chance that a low-risk client who tests positive actually has HIV.

Session 4 was different. This counselor was the only one among all 20 counselors who explained that the proportion of false positives among all positive tests depends on the prevalence of HIV; that is, when low-risk clients test positive, the proportion of false positives can be substantial. Session 4 took place in a public health center in a large city with a population of more than 1,000,000.

Session 4: The Counselor Was a Female Social Worker

Sensitivity?

- Very, very reliable.

- No, not absolutely sure, such a thing doesn't exist in medicine, because it's possible that the virus cannot be identified.

- Close to 100 percent; I don't know exactly.

False positives?

- They exist, but are extremely rare.

- On the order of one tenth of a percent. Probably less. However, in your risk group, compared to high-risk groups, false positives are proportionally more frequent.

- I don't know the exact value.

Prevalence?

- With the contacts you have had, the infection is unlikely.

- Generally, one can't say. In our own institution, out of some 10,000 tests in the last 7 years, there were only 3 or 4 heterosexuals, non-drug addicts, or similar non-risk-group people who tested positive.

Positive predictive value?

- As mentioned, the test is not 100 percent sure. If the test con-
founds the [HIV] antibodies with others, then other methods,
such as repeated tests, do not help. And if someone like you does
not have any real risk, then I could imagine that even 5 to 10 percent of those who receive a positive result will have gotten a false
positive result.

The counselors in Sessions 2 and 3 had dropped remarks suggesting that
prevalence had little value for evaluating Ebert's case. In contrast, the
counselor in Session 4 understood the relationship between prevalence
and the positive predictive value: If a client who is in a group with a low
prevalence tests positive, the danger of a false positive is particularly high.
This counselor was also the only one to explain that false positives cannot
be completely eliminated by repeated testing—for instance, the test may
react to antibodies that it confuses with HIV antibodies. The counselor still
probably overestimated the positive predictive value, but her estimate was
on the right order of magnitude.

What information did the rest of the counselors provide, and how did
they communicate it?

TWENTY COUNSELORS

One physician in a small Bavarian city declined to give the client any information whatsoever concerning the HIV test's sensitivity, specificity, and positive predictive value before the result was obtained, which left us with
responses from 19 counselors. Most of the counselors gave the client realistic
information concerning sensitivity, although 5 incorrectly claimed that false
negatives are impossible except during the window period. The exchange of
test results in Susan's case (Chapter 1) illustrates one source of false negatives; while she got a *false positive*, the person whose result was confused with
hers got a *false negative*. The question concerning the prevalence proved to
be a hard one for the counselors; most could not find the information. Several searched in their files and brochures for an answer, but found only irrel-

evant prevalences, such as the larger number of HIV positives in West Berlin than in East Berlin. "The Wall was the best condom East Berlin had," one counselor joked in desperation after failing to find an answer.

The counselors at the public health centers were not ignorant; on the contrary, several gave lengthy and sophisticated lectures concerning immunodiagnostic techniques, the nature of viruses, antibodies, and proteins, and the pathways of infection. But when it came to explaining Ebert's risk of being infected if he tested positive, most counselors lacked the ability even to estimate, much less to communicate the risks.

To summarize, the principal deficits in the counseling of low-risk clients were:

- *Nontransparent risk communication.* All of the counselors communicated information in probabilities and percentages rather than in a format, such as natural frequencies, that helps their clients (and themselves) attain insight. As a consequence, several counselors did not notice that the numbers they supplied were internally inconsistent. For instance, one counselor told the client that the prevalence of HIV in men like the client was 0.1 percent or slightly higher, and that the sensitivity, specificity, and positive predictive value were each 99.9 percent. However, these figures are impossible, as a frequency representation can quickly demonstrate.

- *Denial of false positives.* The majority of the counselors (13) incorrectly assured the client that false positives never occurred. Counselors had a simple, deterministic explanation for this: Any false positives would be eliminated through repeated testing, that is, through two ELISAs and one Western blot test. One counselor claimed that false positives occurred only in countries such as France, but not in Germany, while others acknowledged that false positives had occurred in the 1980s, but not since. In addition to these 13 counselors, 3 others first claimed that false positives would not occur (for example, Sessions 2 and 3), but then had second thoughts.

- *Failure to understand that the proportion of false positives is higher in low-risk clients.* Only one counselor (Session 4) explained that the lower the prevalence, the higher the proportion of false positives among positive tests. In other words, the ratio of false positives to true positives is particularly high among low-risk people such as Ebert.

- *Illusion of certainty.* Ten of the counselors asserted incorrectly that if a low-risk man tests positive, it is absolutely certain (100 percent) that he is infected with the virus (see Session 1), and 5 others told Ebert that the probability is 99.9 percent or higher (see Session 3). Based on the best figures available, this probability is, in fact, around 50 percent (see Figure 7-1). If Ebert had tested positive and trusted the information provided by one of these 15 counselors, he might indeed have contemplated suicide, as had others before him. Two other counselors successfully avoided answering the question concerning the positive predictive value. Only three counselors estimated this probability to be less then 99.9 percent (all three estimates exceeded 90 percent). Counselors arrived at this prevailing illusion of certainty by one of two routes. Some confused the positive predictive value with the sensitivity. Others assumed that there are no false positives because tests are repeated, which implies that a positive test indicates an infection with absolute certainty.

The lesson of this study is this: First, counselors need to be trained to overcome the illusion of certainty, and second, they need to be taught how to communicate the risks so that clients (and they themselves) understand them. The following model session illustrates how a counselor could communicate the meaning of a positive test transparently.

Model Session
The counselor is trained in communicating risks in natural frequencies.

Sensitivity?

- The test will be positive for about 998 of 1,000 people infected with HIV. Depending on circumstances, such as the specific tests used, this estimate can vary.

False positives?

- About 1 in 10,000. False positives can be reduced by repeated testing (ELISA and Western blot), but not completely eliminated. They are caused by certain medical conditions as well as by laboratory errors.

Prevalence?

- About 1 in 10,000 heterosexual men with low-risk behavior is infected with HIV.

Positive predictive value?

- Think of 10,000 low-risk men like you. One is infected and will test positive with practical certainty. Of the 9,999 noninfected men, 1 will also test positive. Thus we expect that out of 2 men who test positive, only 1 has HIV. This is the situation you would be in if you were to test positive; your chance of having the virus would be about 1 in 2.[21]

Depending on the region and risk behavior, the numbers can be adjusted accordingly. The counselor might add that in the case of a positive result, a second blood sample should be tested. A second sample can rule out certain sources of error, but not all. The same holds for a negative test, as one extreme case illustrates. In the VA Medical Center in Salt Lake City, a man with HIV had 35 negative tests within a 4-year period.[22] This case is unusual because the man had a strain of HIV typical to the United States, but the tests could not detect his antibodies. We cannot expect complete certainty, but we certainly can improve the present state of affairs with medical and psychological research, that is, with medical weapons against the virus and with mind tools to support human reasoning in understanding risks.

Information Leaflets

Do the leaflets and brochures available in public health centers help clients understand what a positive test means when prevalence is low? To answer this question, we analyzed 78 leaflets and brochures available on AIDS and HIV testing in the 20 German public health centers. Some of these were given to Axel Ebert by the counselors.

The strengths and weaknesses of these brochures mirrored those of the counselors. Their strength was to provide plenty of relevant and useful information on how the HIV virus is transmitted and how to live with HIV. Their blind spots concerned what a positive test means and how this meaning depends on the risk behavior of the client. Several brochures mentioned that false negatives and false positives can occur. For instance, in the newsletters edited by the Federal Center for Health Education, the reasonable recommendation is made that people at no known risk who nevertheless test positive should ask for a second test.[23] (One wonders what those counselors who believe that a positive result is definitive will tell such a client.) However, none of the leaflets and brochures gave estimates of how often false negatives and false positives occur. Rather, one promised that antibody tests that identify HIV-1 and HIV-2 infections with *certainty* will be available "in the near future." None explained that the chance of being infected given a positive test strongly depends on risk behavior. The most recent newsletter only stated, "A positive HIV test only means that an infection with the virus has occurred. It does not necessarily mean that you have AIDS now and all is lost."[24] Reading this brochure, one gets the impression that a positive test means being infected with HIV; there is no distinction between high- and low-risk behavior, and no mentioning of false positives. A handbook for counselors, in contrast, correctly explained that the false positive rate is less than 1 in 1,000, but then confused the definitions of sensitivity and specificity.[25] With brochures of such mixed quality, neither counselor nor client is likely to understand the outcomes of HIV tests.

Do Americans receive better information? An analysis of 21 AIDS leaflets distributed at the University of Chicago Hospital, the Howard Brown Memorial Clinic, and other medical institutions in Chicago would suggest not.[26] These leaflets classified various sexual behaviors as safe, risky, or in-between and made recommendations about how to prevent HIV infection. Some mentioned the possibility of false negatives in the window period. Not a single one mentioned the possibility of false positives. For instance, one leaflet "Coping with HIV Disease," distributed by the Illinois Department of Public Health, left no room for uncertainty: "A person who is HIV positive has HIV disease."

The major difference between the American and the German leaflets was that the former did not even mention the possibility of false positives. Not having this basic information makes it hard to understand that the proportion of false positives is particularly high among low-risk clients who test positive.

Should Anything Be Done?

Yes. Counselors need to be cured of their illusion of certainty and trained in communicating risks transparently. This will not prevent AIDS, but it will prevent some of the avoidable consequences. These include the danger that people with false positives will have unprotected sex with people infected with HIV, that they may suffer for months or years believing they are infected, and that they think of committing suicide or actually do commit suicide. As many as 30 percent of people seeking HIV testing at a New York hospital reported having suicidal thoughts during pretest counseling.[27]

Since the first AIDS cases were described in 1981, more manpower and money have been poured into researching HIV than any other disease in history. Little, in contrast, has been done to educate the general public about what an HIV test result means. An educated public that understands what a test result means will be even more crucial the moment self-administered HIV tests become commercially available. The Food and Drug Administration had for years opposed home-based HIV testing because of the lack of face-to-face counseling. In 1996, however, it reversed its position and approved two self-administered test kits. One of the reasons was the hope that home-based kits could reach those estimated 80 percent of people with high-risk behavior who are not willing to visit HIV testing sites.[28] These kits can be purchased by individuals of any age. All one has to do is to prick one's finger, place three drops of blood on a card, mail it anonymously to a laboratory, and call in a week later and punch one's identification number into the phone. If the test is positive, one will be connected to a phone counselor; if negative, to an automatic recording. Given the problems present in face-to-face counseling documented in this chapter, problems arising from telephone counseling are likely to be even more serious.

As I write these lines, a friend of mine notifies me of the case of a young man whose doctor informed him on the phone that he had tested HIV-positive. The man immediately committed suicide—there was no post-test counseling and no second test with a new blood sample. If David from Dallas had committed suicide, too, we would probably never have known that his result was a false positive.

The courtroom oath—"to tell the truth, the whole truth and nothing but the truth"—is applicable only to witnesses. Defense attorneys, prosecutors, and judges don't take this oath—they couldn't! Indeed, it is fair to say the American justice system is built on a foundation of *not* telling the whole truth.

Alan M. Dershowitz, The Best Defense

8

WIFE BATTERING

Los Angeles, USA

The verdict was to be announced at 10:00 A.M. on October 3, 1995. The Los Angeles Police Department was on full alert, and nationwide security measures, on which President Clinton had been briefed, were taken in case of rioting. As the hour approached, long-distance call volume dropped by 50 percent, exercise machines in gyms stood idle, work ceased in factories, and trading volume on the New York Stock Exchange fell by 40 percent. An estimated 100 million people switched on their televisions and radios to hear what the jury had decided in the O. J. Simpson trial. How would 12 jurors, two-thirds of whom were black women, judge a black man accused of murdering his ex-wife, Nicole Brown Simpson, a white woman, and her male companion? At the appointed hour, Judge Lance Ito called the courtroom to order. Then the jury delivered the verdict: not guilty.

Images of college students watching the verdict swept across the world: young black women jumping up, cheering, hugging, and clapping hands; young white women sitting, stunned into silence, chins propped on their hands. The acquittal of the American football star divided people along racial lines. Race trumped gender.

But the trial might as easily have turned on gender as race. The most dangerous piece of evidence facing the Simpson defense threatened to divide the jury along gender rather than racial lines—namely, Simpson's history of spousal abuse. There had been at least one incident in which Simpson treated his wife violently, and numerous other incidents that suggested that he exhibited sexual jealousy and a tendency toward violence. The prosecution argued that a history of spousal abuse reflects a motive to kill and spent the first 10 days of the trial calling witnesses of the 18-year relationship between Simpson and his wife to the stand. As one prosecutor put it, "a slap is a prelude to homicide."[1]

Alan Dershowitz, a renowned Harvard law professor, advised the Simpson defense team. In his best-selling book *Reasonable Doubts: The Criminal Justice System and the O. J. Simpson Case,* Dershowitz explained the team's success in quashing the prosecution's argument that spousal abuse leads to murder. Dershowitz claimed repeatedly that evidence of abuse and battering should not be admissible in a murder trial: "The reality is that a majority of women who are killed are killed by men with whom they have a relationship, *regardless of whether their men previously battered them.* Battery, as such, is not a good independent predictor of murder."[2]

What was the evidence for Dershowitz's claim? The defense told the court that some studies estimated that

> as many as 4 million women are battered annually by husbands and boyfriends [in the United States]. . . . Yet in 1992, according to the FBI Uniform Crime Reports, a total of 913 women were killed by their husbands, and 519 were killed by their boyfriends. In other words, while there were 2½ to 4 million incidents of abuse, there were only 1,432 homicides. Some of these homicides may have occurred after a history of abuse, but obviously most abuse, presumably even most serious abuse, does not end in murder.[3]

From these figures Dershowitz calculated that there is less than 1 homicide per 2,500 incidents of abuse: "we were convinced from the very beginning that the prosecutors' emphasis on what they called 'domestic violence' was a show of weakness. We knew that we could prove, if we had to, that an infinitesimal percentage—certainly fewer than 1 of 2,500—of men who slap or beat their domestic partners go on to murder them." Dershowitz concluded: "There is never any justification for domestic violence. But neither is there any scientifically accepted evidence that domestic abuse—even of the sort attributed to Simpson in the worst-case scenario—is a prelude to murder."[4]

Sounds convincing, doesn't it? In the jargon of the law, Dershowitz argued that evidence of wife battering is more "prejudicial" than "probative." If this were true, acquittals of men like O. J. Simpson, known to have battered his murdered wife on at least one occasion, could be better justified in the future. But Dershowitz's argument is misleading. And it may have misled the court. Given the "scientifically accepted evidence" available, if a woman has been battered by her partner and later murdered, the case against the batterer is in fact fairly strong. Why?

Dershowitz omitted one crucial piece of evidence from his calculation: that Nicole Brown Simpson had been murdered, not just battered. The relevant percentage is not how many men who slap or beat their domestic partners go on to murder them, as Dershowitz would have us believe. Instead, the relevant probability is that of a man murdering his domestic partner given that he battered her *and* that she was murdered.[5] This probability is not 1 in 2,500. What is it?

There are two ways to estimate this probability, just as in the breast cancer and other medical examples in the previous chapters. One is to insert probabilities into Bayes's rule, which can cause confusion in jurors and judges and even in experts who perform such calculations. The other is to present the information in more easily understandable natural frequencies. In the following explanation of the flaw in Dershowitz's argument, I use his estimate that 1 in 2,500 battered women is killed annually by her husband or boyfriend. This corresponds to 40 in 100,000. What we need in addition is the number of battered women who are killed each year by someone other than their partners. Assuming that this number is about the same as for all American women (whether battered or not), the *Uniform Crime Reports for the United States and Its Possessions* (1993) estimated 5 women in 100,000 who are murdered each year.[6] Using frequencies, we can now easily understand how many battered and murdered women have been murdered by their husbands or boyfriends (Figure 8-1).

Think of 100,000 battered women. Within one year, we can expect about 40 to be murdered by their batterers and another 5 to be murdered by someone else. Therefore, 40 out of every 45 murdered and battered women have been killed by their batterers. That is, in only 1 out of 9 cases is the murderer someone other than the batterer.

The frequency tree makes this reasoning transparent. It is like the tree that helped to make the chances of breast cancer after a positive mammogram transparent. In each case, a sample of concrete cases—in this case, battered women—is broken into subclasses, as would occur during natural sampling. Thus, the chances that a batterer actually murdered his partner given that she has been killed is about 8 in 9, or approximately 90 percent. This probability must not be confused with the probability that O. J. Simp-

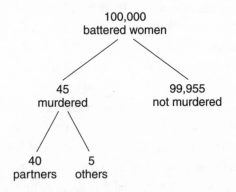

FIGURE 8-1. *Is there a relationship between wife battering and spousal murder in the United States?* The Harvard law professor Alan Dershowitz, who advised the O. J. Simpson defense team, argued that battering does not constitute evidence against the husband (or partner) when his wife has been murdered. If one draws a frequency tree based on Dershowitz's numbers, however, one sees the flaw in the argument. At the time of the trial, evidence suggested that out of every 100,000 battered women, 45 were murdered every year. In 40 of these cases, the murderer was the victim's partner. Thus, wife battering *is* evidence against the partner of a murdered woman.

son is guilty; a jury must take into account much more evidence than battery to convict him beyond a reasonable doubt. But this probability shows that battering is a fairly good predictor of guilt for murder, contrary to Dershowitz's assertions. Evidence of battery is probative, not prejudicial.

I cannot judge whether Dershowitz confused himself with his numbers, or whether he just confused the court, the public, and the readers of his book about the Simpson case. And this may not even be an important question for Dershowitz, given his baseball analogy of the legal system, according to which "most prosecutors (and defense attorneys) are as concerned about their won-lost ratio as any major-league pitcher." In his words, "nobody really wants justice. Winning is 'the only thing' to most participants in the criminal justice system—just as it is to professional athletes."[7] Either way, the moral is the same. Baseball grew up from sandlots and city streets in a culture of working men and farm boys, and professionals today play the statistics as well as playing ball. Little boys are familiar with batting averages, won-lost percentages, and lots of other statistics. The same cannot be said of the court. Many students who spent much of their education avoiding math and statistics become lawyers, and are unfamiliar with conditional probabilities, match probabilities, and other statistical evidence. Representing statistical information in natural frequencies can help courts—and the public—win arguments and gain insight into the actual relationship between wife battering and murder.

Battering: The Larger Context

Unlike cancer and HIV, wife battering is a social disease that is manmade. A look into its history may help us see how this phenomenon might be changed, at least slowly. Knowing how frequent violence is against women, even today in Western democracies, can perhaps help to expose the often hidden problem.

In 1874, Richard Oliver arrived home drunk one morning, threw a cup and a coffee pot on the floor objecting to his wife's cooking, brought in two tree branches and whipped her, and could be stopped only after witnesses intervened. Neither of the branches was as thick as a man's thumb. Nevertheless, Judge J. Settle affirmed that he was guilty of assault and battery, because the "rule of thumb" doctrine was not law in North Carolina. According to this old doctrine, a husband had a right to whip his wife, provided he used a stick no thicker than his thumb. This court case established that, at least in North Carolina "the husband has no right to chastise his wife under any circumstances."[8] Around the same time, Judge Charles Pelham wrote: "The privilege, ancient though it be, to beat her with a stick, to pull her hair, choke her, spit in her face or kick her about the floor, or to inflict upon her like indignation, is not now acknowledged by our law." Some 120 years after Judge Pelham condemned this sort of wife beating, samples of affidavits filed in U.S. courts show that the same violent behaviors are still reported by American women.

The evolutionary psychologists Martin Daly and Margo Wilson argue that most cases of wife battering arise from husbands' jealous and proprietary responses to their wives' real or imagined infidelity or desertion.[9] In a small percentage of these cases, battering "boils over" into murder. (Recall that Nicole Brown Simpson was murdered along with a male companion, whom her husband—were he the murderer—could have viewed as a sexual competitor.) Battered women often report that their partners threaten to kill them. In fact, of all American women murdered every year, some 30 to 40 percent are killed by their intimate partners.[10] A recent national study of hospital emergency rooms found that assaults by intimate partners or former partners accounted for more than 45 percent of women's violence-induced injuries (among all cases where an assailant could be identified).[11] In a random sample of San Francisco women, 1 out of 5 women reported having been beaten. Women receiving welfare assistance seem to be at greater risk than the general population; for instance, 2 out of 3 Massachusetts women on welfare reported having been victims of wife beating. Little research has been conducted among the economically privileged in America, except by journalists. (The Pulitzer Prize–winning Seymour Hersh, for example, claimed that former President Richard Nixon beat his wife Pat and hospitalized her a number of times.)[12]

Physical abuse of women is not restricted to partners. Every year, about 1 in 1,000 American women aged 12 and older is raped.[13] This figure is the reported forcible rape rate; because of the underreporting of rape, the actual number is higher. Estimates are that one in five or six American women has been the victim of a completed rape at some point in her life.[14]

Coercive control over women seems to have existed in all or most cultures and can still be observed. However, this does not mean that abuse cannot be drastically reduced. Adultery laws exemplify one such change. Ancient Egyptians, Syrians, Hebrews, Romans, Spartans, and other Mediterranean cultures defined adultery solely in terms of the marital status of the woman—that is, as an affair between a married woman and a man, married or not. Adultery was seen as a kind of theft from the husband, sexual access to the wife being the commodity stolen. Consistent with this view, until 1974, Texas law (Texas Penal Code 1925, article 1220) allowed a husband to kill a man whom he found in the act of adultery with his wife, without any penalty whatsoever. Texas is an extreme case in being so late to change. In 1852, Austria became the first country to explicitly treat male and female adultery as legally equivalent. In 1996, Austria once again took the lead by passing an exceptional law against wife battering. As a consequence of this law, battered women need not seek refuge in homes for battered women; they need only call the police, who will instantly turn up at the door and take the husband out of the home if the wife wishes it. The police will confiscate his keys to the house and can forbid him from entering the neighborhood for up to 3 months.

Men's susceptibility to violence can shift with changes in the economic and political environment. In the Soviet Union, for example, the number of women killed by their partners was reported to be about 1,600 in 1989, 1,900 in 1990, and 5,300 in 1991.[15] After the Soviet Union's disintegration, the Russian Federation's *National Report* noted that 14,500 women were killed by male partners in 1993 (in Russia only), a figure that rose to 15,000 and then 16,000 in the following two years. The rate of spousal murder in the Russian Federation surpassed that in every Western country, even the high rate in the United States, by an order of magnitude. The worsening economic, political, and social conditions in Russia during the turmoil that followed the fall of the Communist regime—notably, unemployment and many Russian men's increased flight into alcoholism—may have fueled the growing violence.[16]

Wife battering is not just a Russian or an American problem. The fact il-

lustrated in Figure 8-1—that a large majority (8 out of 9) of battered
women who are murdered are killed by their partners—seems to hold in
Western countries generally. Men are also much more likely to kill their
spouses than are women, a fact that also seems to hold in all countries. In
the relatively few cases in which women kill their partners, the women have
typically been abused and beaten by their partners for years. An analysis of
German court cases showed that in four out of five cases in which a woman
was convicted of homicide, she had murdered her partner and that the
majority of these women had been battered by their partners.[17] Rural and
urban Mexican women have reported that most of the violence they experience comes from their partners, whether owing to alcoholism, financial
problems, jealousy, women's true or imagined infidelity, or the birth of a
child of the "wrong" sex (that is, a girl).[18] A recent report on domestic violence in Chile describes the commonly accepted belief in Chilean culture
that men may demonstrate love through violent acts. Together with ma-
chismo and alcoholism, this mind-set makes violence against women and
children a "way of life."[19] Until 1989, Chile's Civil Code legally sanctioned
the disregard of human rights for women, stating that a wife owes obedi-
ence to her husband, who protects her and has authority over her possessions and person.

Why is there wife battering all around the world? And why do many
more men than women kill their partners? Although there are many con-
tributing variables, such as alcoholism, the classic explanation is paternal
uncertainty. Unlike in the majority of mammalian species, in which males
contribute nothing to the upbringing of their offspring, in the human
species males and females cooperate in providing parental care. Fathers
face a problem that mothers do not, and which according to evolutionary
theory is so serious that most mammalian fathers opt out of paternal investment entirely. This problem is cuckoldry. That is, a man has to accept
some degree of uncertainty about whether he actually is the father of his
children. A woman, in contrast, can be certain that she is the mother of her
children (barring an accidental exchange of babies in the hospital). Paternal uncertainty can be reduced by many means, one of which is for a man
to control his partner physically to ensure that she is not consorting with
other men. According to this argument, the cost of male parental invest-

ment brings with it male sexual jealousy, which leads men to use methods ranging from vigilance to violence to controlling sexual access to their mates.

How certain should human fathers be about their paternity? Probably not as certain as is conceivably possible since the mid-1980s, when DNA fingerprinting became available as a highly reliable method for paternity testing. Using DNA fingerprinting, researchers found that 5 to 10 percent of children in Western countries who had been studied have a different biological father from the one they thought they had.[20]

DNA fingerprinting has understandably played a critical role in paternity suits, and its potential and limitations in criminal cases were also highlighted in the O. J. Simpson trial. Interpreting a DNA match, like seeing through Dershowitz's denial of the connection between wife battering and murder, requires clear statistical thinking. In Chapter 10, we will take a closer look at DNA technology.

> ... the theory of probabilities is at bottom only common sense reduced to calculus ...

Pierre-Simon, Marquis de Laplace,
A Philosophical Essay on Probabilities

9

EXPERTS ON TRIAL

Los Angeles, USA

On June 18, 1964, Juanita Brooks walked down an alley toward her home in the San Pedro area of Los Angeles. With one hand she held her cane, and with the other she pulled behind her a wicker carryall containing groceries, with her purse on top of the packages. Suddenly, she was pushed to the ground by a person she neither saw nor had heard approach. She managed to look up to see a young woman running from the scene. Mrs. Brooks's purse, which contained between $35 and $40, was missing. A witness who lived at the end of the alley reported that the running woman was blond, had a ponytail, was dressed in dark clothing, and fled from the scene in a yellow car that was driven by a black man with a beard and a mustache.[1]

Armed with this description, the police arrested a couple, Janet and Malcolm Collins, who fit the description. The couple had married just two weeks before, at which time they had only $12, a portion of which was spent on a trip to Tijuana. Since the marriage ceremony, Malcolm had not worked, and Janet's earnings as a housemaid were not more than $12 a week. During the 7-day trial, the prosecution ran into difficulties. Although he identified Malcolm Collins in the courtroom as the driver of the getaway car, the witness had admitted at a preliminary hearing that he was uncertain of his identification of a beardless Mr. Collins in a police lineup. Mrs. Brooks could not identify either defendant.

To make a case, the prosecutor in *People v. Collins* called a mathematics instructor from a state college to the stand as an expert witness. The prosecutor provided a chart similar to Table 9-1 to answer the following question: What is the chance that Mr. and Mrs. Collins are innocent given that they match the descriptions of the perpetrators on all six characteristics? The expert witness testified that the probability of a combination of characteristics, or their *joint probability,* is given by the product of their individual probabilities. Then the prosecutor provided estimates of the probabilities of each of the six characteristics and multiplied them to compute what he claimed to be the probability that a randomly selected couple would have all six characteristics; his answer was 1 in 12 million.[2] Based on this calculation, the prosecutor concluded that the chance that the defendants were innocent was only 1 in 12 million. He added that this estimate was conservative and that, in reality, "the chances of anyone else besides these defendants being there . . . is something like 1 in a billion."[3] The jury later convicted the Collinses of second-degree robbery.

The defense appealed the verdict, and the California Supreme Court re-

---

TABLE 9-1. *Probabilities provided by the prosecution in* **People v. Collins.** These probabilities are estimated relative frequencies, such as that 1 out of 3 girls have blond hair and 1 out of 10 girls have a ponytail. (Data from Koehler, 1997, p. 215.)

Evidence	Probability
Girl with blond hair	1/3
Girl with ponytail	1/10
Partly yellow automobile	1/10
Man with mustache	1/4
Negro man with beard	1/10
Interracial couple in car	1/1,000

versed the conviction on four grounds. First, the probabilities in the prosecutor's chart lacked evidentiary foundation; they were merely estimates. Second, multiplying the six probabilities requires assuming that the six characteristics are independent, for which there was insufficient proof. Beards and mustaches, for instance, are not independent: a man who has a beard is more likely than a randomly drawn man to have a mustache. Third, the prosecutor's calculation assumed that the six characteristics were certain, ignoring the possibility that the perpetrators were disguised or that the witnesses inaccurately reported one or more of the perpetrators' characteristics. For instance, the report that the female perpetrator had a ponytail was actually uncertain: the victim was not able to state whether she had one—although the victim had observed her as she ran away—whereas the witness was sure she did. Or, the female perpetrator might have been a light-skinned black woman with bleached hair rather than a white woman. Fourth, and most important, there was a fundamental flaw in the prosecution's reasoning. The prosecution inferred that the probability of observing all six characteristics in a randomly drawn couple is the probability that Mr. and Mrs. Collins were innocent. This error is known as the *prosecutor's fallacy*.[4]

To understand this fallacy, we need to separate two questions. First, what is the probability that an individual (or couple) will match the perpetrator on all known characteristics? Second, what is the probability that an individual (or couple) is guilty, given that this individual matches the perpetrator on all known characteristics? The prosecutor's fallacy refers to the conflation of these two probabilities:

$$p(\text{match}) \text{ is mistaken for } p(\text{not guilty}|\text{match})$$
Prosecutor's fallacy

In words, the prosecutor's fallacy is to reason that the probability of a random match is the same as the probability that the defendant is not guilty, or equivalently, that the *guilt probability* is 1 minus the probability of a random match. For instance, assume that $p(\text{match})$ is 1 in 1,000. The person who commits the fallacy reasons that, therefore, the chances that the defendant is not guilty are 1 in 1,000, or, equivalently, that the chances that the defendant is guilty are 999 in 1,000. In fact, however, these two probabilities are not the same.

The probability p(match) is a *random match probability* of a trait or combination of traits (for example, ponytail) in a defined population (for example, all people in the United States). The probability p(not guilty| match), in contrast, is the probability that the defendant is not guilty given a match. The confusion between these two probabilities can be seen easily if we use a trait that matches frequently such as being male if we think about the next U.S. president. The probability p(male) that a randomly drawn American is male is about 50 percent. But this probability is not the same as the probability that any random person will be the next U.S. president if he is male, p(president|male); the vast majority of men never become president.

The attorney and social psychologist William Thompson and his student Edward Schumann seem to have coined the term "prosecutor's fallacy." They used it to describe an experienced deputy district attorney's argument that if a defendant and perpetrator match on a blood type found in 10 percent of the population, there is a 10 percent probability that the defendant would have this blood type if he were innocent, and therefore, that there is a 90 percent probability that the defendant is guilty.[5] If you find this argument confusing, that is because it is a confused argument: A random match probability does not determine a guilt probability. As we have seen in earlier chapters, the prosecutor's fallacy is not specific to prosecutors. The same impeded inferences can occur when other experts reason with probabilities. The prosecutor's fallacy is related to the reasoning of several of the AIDS counselors in Chapter 7 who confused the sensitivity of the HIV test with its positive predictive value, when, in fact,[6]

p(positive test|HIV) is not the same as p(HIV|positive test)

The prosecutor's fallacy is also related to the reasoning of those physicians in Chapter 5 who confused the sensitivity of mammography with the probability that a woman actually has breast cancer if she tests positive, when, in fact,

p(positive test|breast cancer) is not the same as p(breast cancer|positive test)

Experts who commit the prosecutor's fallacy often use the phrase "someone other than the defendant":

> *Because the probability of a match between the defendant and the evidence sample is 1 in 10,000, the probability that someone other than the defendant is guilty is only 1 in 10,000.*

How can this form of innumeracy be avoided? A simple solution would be for the courts to require that evidence be presented in frequencies rather than single-event probabilities. In the case of the Collinses, the confusing probability statement is this:

> "The probability of the defendant matching on these six characteristics is 1 in 12 million."

The probability statement looks like bad news for the defendant. Transformed into the more transparent language of frequencies, it draws attention to other possible suspects:

> "Out of every 12 million couples, we expect that 1 couple shows these six characteristics."

Here one immediately asks how many couples are out there who could have committed the crime. Frequencies make it apparent that we need to know the number of couples in the relevant reference population to estimate the chance that the Collinses are innocent. For instance, if the reference population was all couples in California, and there were about 24 million couples in California, then the natural frequency statement reveals that we can expect two couples to have all six characteristics. This would mean the chance that the Collinses are innocent is 1 in 2, not 1 in 12 million.[7] In the appeal, the Supreme Court of California made a similar calculation, concluding that convicting the Collinses was like convicting person

X on the grounds that a witness saw either X or X's twin commit the crime. The report continues with a telling remark about the state of statistical insight in the courtroom: "Again, few defense attorneys, and certainly few jurors, could be expected to comprehend this basic flaw in the prosecution's analysis."[8]

People v. Collins is one in a long line of legal cases in which experts, not just laypeople, were confused by probabilities. In the notorious case in late nineteenth-century France, for example, the conviction of Dreyfus for espionage was ultimately reversed and the statistical arguments that had supported it were discredited, just as in the Collins case.[9] Thus, expert testimony using probabilities, and the confusion it can engender, has played a role in the courtroom for more than a century now. Yet it is still common practice in today's courts to present statistical arguments in probabilities rather than in frequencies. This practice makes it difficult for people to spot flawed arguments.

Wuppertal, Germany

On a summer evening in Wuppertal, a city in the industrial belt of Germany, a 40-year-old painter took a walk in the woods with his 37-year-old wife. Suddenly, they were attacked by a stranger who shot the man three times in the throat and the chest. The man fell to the ground. Then the stranger attempted to rape the woman. When she defended herself and, unexpectedly, her husband rose to his feet to help her, the stranger shot her twice in the head and fled. The man survived the attack; his wife did not. Three days later, a forest ranger discovered, 20 kilometers from the scene of the crime, a car belonging to a 25-year-old chimney sweep who used to spend his weekends in the same woods. The husband at first thought he recognized the chimney sweep in a photo, became less certain after seeing him in person, and later leaned toward believing another suspect to be the murderer. But when the other suspect was proven innocent, the prosecution put the chimney sweep on trial. The chimney sweep, who had no previous convictions, pled not guilty.

Among the pieces of evidence against the defendant was the blood found under the fingernails of the murdered woman, which matched the defendant's. At the trial, a university lecturer testified that 17.3 percent of Germans share that blood type. A second piece of evidence was the blood found on the chimney sweep's boots, which matched the murdered woman's. The expert testified that 15.7 percent of Germans share this blood type. Multiplying these two probabilities gives a joint probability of 2.7 percent that these two matches would occur by chance. Therefore, the expert witness concluded that the probability is 97.3 percent that the chimney sweep was the murderer.[10]

To see why the conclusion of the expert witness does not hold water, let us assume that any of roughly the 100,000 men in Wuppertal could have committed the crime. One of them, the murderer, will show both matches with practical certainty (unless the samples are confused with other samples in the laboratory, or other errors occur). Out of the 99,999 other residents, we can expect some 2,700 (or 2.7 percent) also to show these two matches.[11] Thus, the probability of the defendant being the murderer given the two matches is not 97.3 percent, as the expert witness testified, but 1 in 2,700, that is, less than one-tenth of 1 percent.

A second kind of evidence was textile fibers found on the clothes of both the victim and the chimney sweep. Based on the match between these fibers, a second expert witness, from the State Crime Department, used the same reasoning—that is, he committed the prosecutor's fallacy— in order to compute a comparably high probability that the chimney sweep was the murderer. These expert calculations collapsed when the court found conclusive evidence that the defendant was in his hometown, 100 kilometers away from the scene, at the time of the crime.

The assumption that 100,000 men could have committed the crime was made here strictly for the purpose of illustration. If one has independent evidence that limits the population of potential suspects, as in the murder case considered in Chapter 10, the estimate can be changed accordingly. In the Wuppertal case, such evidence was not available. The assumption of a certain population size generally remains an assumption. By taking upper and lower limits on possible population sizes (for example, all men in Europe versus all men in Wuppertal), however, we can derive upper and lower bounds on the probability that a defendant is the source of a forensic sample. In any case, the method of thinking of a concrete population and breaking it down into natural frequencies can provide us with ballpark estimates that prevent the prosecutor's fallacy.

The reasoning of the two expert witnesses in this case was the same as that of the prosecutor in *People v. Collins.* It is no coincidence that this type of clouded thinking is known as the prosecutor's fallacy rather than the defense attorney's fallacy. This is because the fallacy usually involves inflating the probability that the defendant is guilty of a crime. For instance, in the Collins case, the probability of a match was said to be 1 in 12 million, which, via the prosecutor's fallacy, was equated with the probability of another couple being guilty of the crime; this left the Collinses looking very guilty indeed. Thanks to the prosecutor's fallacy, match probabilities—which are usually very small—miraculously turn into huge guilt probabilities.

Some legal scholars have argued that statistics should be excluded from court because they are easily manipulated and difficult to understand.[12] I believe that statistical evidence should not be excluded, but that professionals working in criminal law should be provided with tools to understand statistical information. In law school, students should be taught using cases such as the two in this chapter and then asked to imagine playing the two following roles. Assume you are a prosecutor and intend to confuse the court; how would you present the evidence? Now assume you are a defense attorney who hears the prosecutor's probability statement; how would you explain, in simple words that a juror can understand, what is wrong with this argument?

Prosecutors, defense attorneys, and judges need to realize that the important question is not only whether evidence is admissible and true, but also whether it is presented in a way that clouds thinking or facilitates insight. Unless courts rule how statistical information should be presented in a trial, we can expect the same errors to be repeated again and again. DNA fingerprinting is a case in point. But it is also a point of hope because if there is one development that has forced the legal profession to learn statistical thinking, it is the admissibility of DNA evidence. Appropriately understood, this new technology promotes the cause of justice.

The Law and Uncertainty

Many students who spent much of their life avoiding statistics and psychology become lawyers. Out of some 175 accredited law schools in the United States, only one requires a course in basic statistics or research methods.[13] When I was a visiting professor at one of the top schools of law in the United States, I was impressed with how smart and rhetorically apt the students were and yet how ignorant they were of basic statistical principles. The same students who excelled in critical thinking could not evaluate whether a conclusion drawn from statistical evidence was correct or incorrect. Similarly, most students were equally ignorant of psychology, including how to communicate statistical information in a way that other minds understand it. These students, however, soon realized that both statistics and psychology are relevant to their careers, since many legal issues are decided on the basis of uncertain evidence. Polygraphs, fiber analysis, hair analysis, DNA fingerprinting, blood-type analysis, and handwriting identification are all techniques that provide evidence that is uncertain and must be evaluated. The courts tend to call in expert witnesses, but as we have seen, the experts that courts select can have clouded minds themselves.

How can this situation be changed? The first step is that legislators, administrators, and law professors recognize that there is a problem. The second step requires looking across the disciplinary boundaries to knowledge of what is helpful in solving the problem and including it as part of law curricula and guidelines for court procedures. Until such a program is instituted, students can try to implement a self-help program, open to any student of law and other sufferers of this problem. As one legal scholar proposed, why not "Innumerates Anonymous"?[14]

Behavioral research should be carried out . . . to assess how well various ways of presenting expert testimony on DNA can reduce . . . misunderstandings.

National Research Council,
The Evaluation of Forensic DNA Evidence

10

DNA FINGERPRINTING

Oldenburg, Germany

In April of 1998, more than 15,000 men between the ages of 18 and 30 went to public schools in Oldenburg, Germany, to have their DNA fingerprints taken. These men, who came from a number of small towns in the area, took this screening very seriously. All were volunteers, and some even had postponed their Easter vacation to participate. The procedure was fast and painless: their DNA was extracted from saliva samples collected using mouth swabs. The cost of this, the largest DNA screening ever conducted in Germany, was estimated at more than $2 million. What had impelled these men to participate was a dreadful crime. One evening in March of the same year, 11-year-old Christina Nytsch ("Nelly") did not come home from swimming at the local public pool. Her alarmed parents called the police, who, after five days of searching, found the child raped and stabbed in the woods.

The DNA traces found on Christina matched those found on another girl who had been raped two years earlier in a neighboring town. That girl had survived. From her testimony, the police were able to make a rough estimate of the perpetrator's age and infer that he lived in the rural region where Christina had lived. Thus, they hypothesized that the population group from which the perpetrator came was that of men between 18 and 30 in that region. The police decided to screen all of them. However, they could not legally force any man to participate; instead, they counted on the locals' moral outrage at the rape and murder of Christina and on the peer pressure that would be brought to bear on anyone who refused to participate.

The police were right. On Good Friday, an unemployed 30-year-old man, married and the father of three small children, was asked by two acquaintances to join them in going to the screening site. The man went. He may have felt he had little choice unless he wanted to be singled out as a suspicious case. After his DNA profile was found to match the one found at the scene of the crime, he confessed to Christina's rape and murder. He also confessed to the rape of the other girl two years earlier. In the case of Christina, a unique combination of social peer pressure and a new technology helped the police to identify a murderer.[1]

In this chapter, I examine the chain of inference from an observed DNA match to a defendant's guilt. This chain involves statistical reasoning that often goes awry. I begin with a brief history of fingerprinting and DNA evidence.

DNA Evidence

Sir Francis Galton (1822–1911), a first cousin of Charles Darwin's, was a romantic figure in the history of statistics and perhaps the last gentleman scientist.[2] He was a man of ideas and a pioneer in diverse fields, including meteorology, the study of heredity, and the measurement of intelligence and the efficacy of prayer. For instance, he collected data to test whether prayer affords any advantage to its intended beneficiaries, and found none: sovereigns for whom whole nations prayed lived no longer than other members of the prosperous classes, nor did clergymen. He also pioneered the use of fingerprints for personal identification. The Galton-Henry system of fingerprint classification was introduced at Scotland Yard in 1901 and quickly adopted in English-speaking countries. In 1924, two large fingerprint collections were consolidated to form the basis of the FBI Identification Division's current database. By the late twentieth century, the division's files contained the fingerprints of more than 90 million people.

In the early 1950s, the chemical basis of heredity was discovered: a sequence of four bases (represented by the letters A, C, G, and T) called DNA (deoxyribonucleic acid). In the 1980s, a new "fingerprinting" technique— *DNA fingerprinting,* or somewhat less fashionably, *DNA profiling*—was developed by Sir Alec Jeffreys of the University of Leicester in England. Jeffreys developed a method for viewing fragments of DNA extracted from blood or other biological samples. The fragments that interested Jeffreys were noncoding regions of the human genome, that is, DNA that has no known function in the cellular production of protein.[3] Because noncoding DNA is subject to less selection pressure than coding DNA, it shows a higher variability among individuals. This higher variability makes it a better tool for discriminating among people, leading to the development of a tool with enormous potential. The fingerprint analogy is not perfect; for example, identical twins have identical DNA, but not identical fingerprints. However, unlike fingerprints, DNA can be extracted from all kinds of biological traces, such as hair, saliva, and blood, and it lasts much longer.

DNA fingerprinting was widely hailed as the greatest breakthrough in forensic science in the twentieth century because of its evident applicability to criminal cases and paternity suits. Because it is not possible to observe DNA directly using instruments such as high-powered microscopes, DNA fingerprinting typically involves the following indirect method. Forensic laboratories break the long DNA strands into fragments by adding an enzyme to the sample; the fragments are then separated by electrophoresis, plotted onto a nylon membrane, and exposed to X-ray film. This process produces a series of fuzzy lines or bands called an *autoradiogram,* which resembles a supermarket bar code. This pattern is the DNA profile. If the bands in the DNA profile line up within a specified error range with those of a particular sample, a "match" is declared.[4] Computer-based data banks with DNA fingerprints have been, or are in the process of being, installed all over the world. For instance, one month after the murder of Christina, the German Federal Crime Bureau established a DNA data bank to facilitate the capture of sexual offenders.

DNA fingerprinting has a vast potential for identifying criminals. For instance, in the investigation of the Unabomber, the FBI obtained a DNA sample from the saliva left on the stamp affixed to one of the bomber's letters and was able to demonstrate a match with Theodore Kaczynski's DNA.[5]

DNA evidence is a valuable tool not only for identifying criminals, but also for preventing false convictions. The defense lawyers Peter Neufeld and Barry Scheck, celebrated for their work on behalf of the O. J. Simpson defense team, run a program, the Innocence Project, the mission of which is the exoneration of people who have been wrongly convicted of crimes. With the help of New York City law student volunteers, they have used DNA fingerprinting to check convictions and have helped overturn the convictions of several dozen wrongly incarcerated people.[6] Similarly, the London Metropolitan Police Forensic Science Laboratory has found that about 20 percent of suspects in rape cases can be excluded by DNA fingerprinting.[7] The possibilities afforded by DNA technology have since spread into the world of science fiction, as in the best-selling novel *Jurassic Park* and the movie based on it, in which extinct dinosaur species are resurrected with the help of as yet fictional biotechnology.

DNA fingerprinting was introduced into U.S. criminal proceedings in 1986. Ten years later, in Britain, the first court case was reported in which the prosecution relied solely on DNA evidence and charged the defendant with rape, despite contrary evidence—for example, that he was considerably older than the description of the assailant by the victim, that the victim did not pick him out of a line-up, and that he had an alibi supported by his girlfriend. The number of suspects who are prosecuted on the basis of DNA evidence alone will possibly increase in the future as a consequence of the creation of national DNA databases, which allow the police to search the computer rather than for further evidence at the scene of the crime.[8]

DNA fingerprinting was initially received by the courts and the media as a nearly foolproof means of identifying criminals who left biological traces at crime scenes, as well as men implicated in paternity suits. However, like all new technologies, DNA fingerprinting depends on psychology to be understood. That is, a great technology can be foiled by a lack of understanding of the uncertainties involved. Two reports by the National Research

Council have addressed the technical and psychological challenges posed by DNA fingerprinting. One of the recommendations in a report in 1996 was that "behavioral research should be carried out to identify any conditions that might cause a trier of facts to misinterpret evidence on DNA profiling and to assess how well various ways of presenting expert testimony on DNA can reduce any such misunderstandings."[9] I hope that this chapter helps to serve that goal.

The Chain of Inference

One would think that a match between a defendant's DNA and the DNA found at a crime scene would prove that the defendant is the source of the trace. But a reported match does not guarantee that the defendant is guilty of the crime or even that he or she is the source of the trace. The chain of uncertain inference from a DNA match to the guilt of a specific person is depicted in Figure 10-1. First, a reported match may not be a true match because of laboratory errors, whether human or technical, that produce false positives, just as in the case of HIV testing. Second, a defendant who provides a true match may not be the source of the trace if the match is coincidental; even rare DNA patterns can occur in more than one person, particularly in biological relatives. Third, a defendant who is truly the source of a DNA trace may not have been present at the crime scene if the real perpetrator or someone else deliberately, or unintentionally for that matter, transferred the defendant's biological material to the scene. In the O. J. Simpson case, for instance, the defense forcefully alleged that some of the blood evidence at the murder scene was introduced by the police. The defense's success in cutting the inferential chain between "Source" and "Present at crime scene" was critical to Simpson's acquittal. Finally, a defendant who had been present at the crime scene may not be guilty; he may have left the trace before or after the crime was committed.

Like the physicians, AIDS counselors, and expert witnesses we have met in previous chapters, attorneys and DNA experts tend to talk about the uncertainties involved in DNA fingerprinting in terms of probabilities. As described in the preceding chapter, the *random match probability* is the

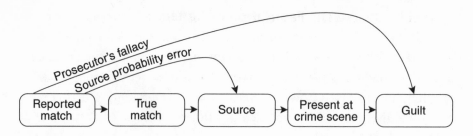

FIGURE 10-1. *The chain of uncertain inference in DNA typing: from reported match to guilt.*
(After Koehler, Chia & Lindsay, 1995.)

probability that a person randomly selected from a population would match the trace evidence as closely as the suspect. The *source probability,* in contrast, is the probability that the suspect is actually the source of the recovered trace evidence. The *guilt probability* is yet something else, because, as mentioned above, even if the suspect is the source of the trace, he may not have committed the crime. The confusion of the random match probability with the other two probabilities leads to two errors. The *source probability error* occurs when someone overlooks the first two steps of the inferential chain and wrongly infers the source probability directly from the random match probability, that is:

$$p(\text{match}) \text{ is mistaken for } p(\text{not source}|\text{match})$$
Source probability error

For instance, if the random match probability is 1 in 100,000, then the source probability error occurs when someone infers that this is the probability that the defendant is not the source, or, equivalently, that the probability that the defendant is the source would be 99,999 out of 100,000. The prosecutor's fallacy, introduced in the previous chapter, involves jumping from the first element of the chain to the last, that is:

$$p(\text{match}) \text{ is mistaken for } p(\text{not guilty}|\text{match})$$
Prosecutor's fallacy

The Fabrication of Certainty

Let us have a closer look at the first step in the inferential chain, "Reported match." If a match is declared, the only error that can be introduced at that step in the inferential process is a false positive result, that is, a match is reported although no true match exists. False positives do occur; unfortunately, it is hard to estimate exactly how often. One reason is that DNA laboratories prefer to conduct in-house proficiency tests rather than to submit to independent external tests of their accuracy.

The FBI, for instance, has fought hard to prevent outsiders from seeing the results of such in-house proficiency tests. The attorney William Thompson reports that the "defense counsel were able to gain access to the FBI's internal proficiency test results only after protracted litigation to overcome the FBI's insistence that these data were protected by a 'self-critical evaluation privilege.' FBI memoranda . . . show that John W. Hicks, the FBI's Assistant Director in Charge of the Laboratory Division, in 1990 unsuccessfully sought authorization to destroy the FBI's DNA proficiency testing data at a time when the FBI was resisting efforts by several defense counsel (including myself) to obtain discovery of those very data."[10] Hicks claimed that "there were no false positive or false negative results in any of the FBI's proficiency tests."[11]

In the few cases in which forensic laboratories agreed to external proficiency testing, the tests were rarely blind, that is, the laboratories and technicians were aware that they were being tested. In the first blind test reported in the literature, three major commercial laboratories were each sent 50 DNA samples. Two of the three declared one false match; in a second test one year later, one of the same three laboratories declared a false match.[12] From external tests conducted by the California Association of Crime Laboratory Directors, the Collaborative Testing Services, and other agencies, the psychologist Jonathan Koehler and his colleagues estimated the false positive rate of DNA fingerprinting to be on the order of 1 in 100.[13] Cellmark Diagnostics, one of the laboratories that found matches between O. J. Simpson's DNA and DNA extracted from a recovered blood stain at the murder scene, reported its own false positive rate to the Simpson defense as roughly 1 in 200.[14]

What causes false positives? The reasons can be either technical, such as enzyme failures that produce misleading DNA banding patterns, or human, such as the contamination or mislabeling of samples and errors in pattern interpretation. There is a need for more public data on false positives and negatives in DNA fingerprinting.

In court, several forensic scientists have been reluctant to acknowledge the possibility of false positives. Excerpted here is an illustrative exchange between an attorney and a forensic expert in a 1991 trial in Texas:

Attorney: Now, you're telling us that you can only get a result or no result; is that correct?

Expert: That's correct.

Attorney: And you couldn't get a false positive?

Expert: There's nothing like a false positive in this, no.

Attorney: How about if you use the wrong sample?

Expert: If you use the wrong sample?

(Attorney nods head.)

Expert: You either get a result, or you don't get a result. There's no false positives.[15]

Here the expert witness repeats his claim that there are no false positives even after being asked about the possibility of an inarguable human error (using the wrong sample). This expert is not alone in promising absolute certainty. American expert witnesses have testified that "there have never been any false positives made with DNA testing at this point"; that "a false positive finding [is] impossible because if the procedures were not correctly followed, no match could be obtained"; and that the accuracy rate is 100 percent and DNA fingerprinting is "failsafe."[16] In Germany, the president of the Society for Forensic Medicine is reported to have publicly asserted that a DNA match "identifies a perpetrator with 100 percent certainty."[17] In other cases, forensic scientists have tried to claim certainty by excluding human error from consideration and reducing the meaning of a false positive to technical error. But this amounts to an illusion of certainty; both technical and human errors contribute to false positives.

Clouded Thinking

Clouded thinking means drawing incorrect inferences from statistics without noticing. The source probability error is one form. Here, the random match probability is confused with the probability that the defendant is the source of the evidence.

IN THE PRESS

The front page of *The Boston Globe* on July 5, 1995, reported on the "big numbers" in the O. J. Simpson trial. The following three paragraphs show how the journalist switches back and forth between confusing random match probabilities and source probabilities, unable to keep them straight:

> As it prepares to wind up its case this week, the prosecution in O. J. Simpson's trial wants jurors to remember the big numbers: The odds are 1 in 170 million, for instance, that someone other than the defendant dripped blood on the scene of Nicole Brown Simpson's murder.
>
> But even figures that sound less stunning—9 percent of all Americans wear size 12 shoes; 1 in 200 people share the "genetic markers" in the assailant's blood; and the odds are 1 in 1,600 that the blood on the killer's glove came from someone other than Simpson—point in the same direction when they're combined.
>
> The chance of anyone fitting all three descriptions is 1 in 3.5 million, and it soars to astronomical levels when factoring in other statistics.

The first paragraph provides a version of the source probability error: The author confuses a random match probability (1 in 170 million) with the source probability, $p(\text{not source}|\text{match})$, that someone other than the defendant dripped the blood, when, in fact,

$p(\text{match})$ is not the same as $p(\text{not source}|\text{match})$.

In the second paragraph, the first two figures are correctly presented as random match probabilities, but the third is phrased in the form of odds (the ratio of two probabilities) and is mistaken as a source probability. In the third paragraph, the author switches back to referring correctly to a random match probability.

IN THE COURTROOM

It is not only the press that is easily confused by probabilities. In the courtroom, confusion often emerges when experts respond to attorneys' questions. For instance, in a 1992 trial for aggravated sexual assault in Texas, *State v. Glover,* the prosecutor questioned the expert witness on direct examination. (In the United States, a lawyer first questions his or her own witness on direct examination; then the other side questions the witness on cross-examination; then the first side is entitled to question the witness again, and then the other side may do so again.) The attorney first asked the expert about a random match probability and then restated this value as a source probability, and the expert corroborated both statements:

Attorney: And are you able to compile all four of those probabilities and determine what is the likelihood of the DNA found in Billy Glover just randomly occurring in some other DNA sample?

Expert: Yes.

Attorney: What is the likelihood of that?

Expert: The way that is done is to multiply each one of those four numbers that I mentioned before together, because each one is separate and independent, and the final number comes out as one in about 18 billion.

Attorney: So the likelihood that DNA belongs to someone other than Billy Glover is one in 18 billion?

Expert: That is correct.[18]

Here again, a random match probability is confused with a source probability. The source probability error has been documented, not only in Texas, but also in other states. For instance, a court transcript from a 1991 California trial involving murder and attempted robbery makes the statement "that the frequency of that DNA banding pattern in the Hispanic population is approximately 1 in 6 billion . . . meant that the chance that anyone else but the appellant left the unknown hairs at the scene of the crime is 6 billion to 1." A transcript from a Kansas trial for rape and murder states that "according to the State's three experts, there was more than a 99 percent probability that Smith was a contributor of the semen found in the swab."[19] In these cases, random match probabilities were erroneously interpreted as source probabilities.

Clouded thinking does not always prevail, however. In Britain, the Court of Appeal quashed the conviction of Andrew Deen for rape because the forensic scientist had committed the source probability error. The scientist described the match probability as the "probability of the semen having originated from someone other than Andrew Deen" and concluded that "the likelihood of this [the source] being any other man but Andrew Deen is 1 in 3 million."[20]

Isn't One in a Billion Enough?

But what if the random match probability is as low as 1 in 57 billion, perhaps the most widely reported figure in the O. J. Simpson case and a figure with a denominator larger than the total population of the earth? Isn't that chance small enough? It is true that the discrepancy between the random match probability and the source probability becomes less significant with smaller random match probabilities. However, there are two reasons a matching suspect might not be the source: one is a coincidental match, the likelihood of which decreases with smaller random match probabilities; the other is a false positive.

False positives play a role in the first link of the chain of inference shown in Figure 10-1, from "Reported match" to "True match," whereas the random match probability plays a role in the second link, from "True match" to "Source." When a false positive occurs in the first link, the second link doesn't matter—the chain is already broken. Specifically, when the probability of a false positive is many orders of magnitude larger than the random match probability, it makes little difference whether the latter is 1 in a million or 1 in a trillion.[21] For instance, if the false positive rate is around 1 in 200, as Cellmark, one of the labs involved in the O. J. Simpson trial, reported, we would expect one erroneous match in each of 200 cases where the defendant is innocent. If a match is false, the random match probability does not matter, however small it is. The match is still false.

One might think that the problem of miscommunicating risks could be solved if attorneys, expert witnesses, and judges understood the relevant probabilities. But even when the prosecution carefully avoids the confusion of the random match and source probabilities, untrained jurors may still confuse them. In the words of one juror, "You can't really argue with science." The DNA evidence "was very conclusive the way the experts explained it."[22]

From Innumeracy to Insight

Does mental defogging work with legal minds as well as with medical minds? To answer this question, Sam Lindsey, Ralph Hertwig, and I asked 127 advanced law students and 27 professionals, mostly lecturers and professors of law at the Free University of Berlin, to evaluate two criminal case files.[23] The files were nearly identical to those in two real rape and murder cases in Germany; the only changes made were those necessary for performing the experimental manipulation and protecting anonymity. In both cases a match was reported between the DNA of the defendant and a trace on the victim, but aside from the match, there was little reason to suspect that the defendant was the perpetrator. Situations in which DNA evidence is the only, or at least the major, evidence against a defendant are likely to occur more and more often. In 1996, the German Federal Constitutional Court ruled that in serious crimes, a judge may order blood samples for DNA analysis even if there is no strong suspicion against the person from whom it is taken. Thus, screening of people for DNA is permissible and does not constitute a violation of civil rights.

Can law students and legal scholars come to understand the uncertainties involved by using frequency representations? Does the representation affect their verdicts, that is, their judgment of whether or not the defendant is guilty? Half of the law students and professionals got the relevant information in probabilities, the other half in natural frequencies. The authentic nature of the case files was highly motivating to the participants, and they spent, on average, more than an hour and a half reading and evaluating the two cases. Since criminal case files are long and detailed, I summarize below the relevant passages for one of the two rape-and-murder cases. As mentioned above, aside from the DNA match there was little reason to suspect that the defendant was the perpetrator.

Conditional Probabilities

The expert witness testifies that there are about 10 million men who could have been the perpetrator. The probability of a randomly selected man having a DNA profile, that is identical with the trace recovered from the crime scene is approximately 0.0001 percent. If a man has this DNA profile, it is practically certain that a DNA analysis shows a match. If a man does not have this DNA profile, current DNA technology leads to a reported match with a probability of only 0.001 percent.

A match between the DNA of the defendant and the traces on the victim has been reported.

Question 1. *What is the probability that the reported match is a true match, that is, that the person actually has this DNA profile?*

Question 2. *What is the probability that the person is the source of the trace?*

Question 3. *Please render your verdict for this case: Guilty or not guilty?*

The three questions concern three steps in the chain of inference from reported match to guilt (see Figure 10-1). The first question is about the probability that the reported match is a true match, the second question is about the source probability, and the third about the guilt of the defendant.

Natural Frequencies

The expert witness testifies that there are about 10 million men who could have been the perpetrator. Approximately 10 of these men have a DNA profile that is identical with the trace recovered from the crime scene. If a man has this DNA profile, it is practically certain that a DNA analysis shows a match. Among the men who do not have this DNA profile, current DNA technology leads to a reported match in only 100 cases out of 10 million.

A match between the DNA of the defendant and the traces on the victim has been reported.

Question 1. How many of the men with a reported match actually do have a true match, that is, that the person actually has this DNA profile?

Question 2. How many of the men with a reported match are actually the source of the trace?

Question 3. Please render your verdict for this case: Guilty or not guilty?

When the information was in probabilities, the legal students and experts were lost. Consider the inference from a reported match to a true match (Question 1). The problem here is false positives. Only very few could figure out what to conclude from a match probability of 0.0001 percent and a false positive rate of 0.001 percent. As Figure 10-2 shows, only about 1 percent of the students and 10 percent of the professionals were skilled enough to do this. When the information was in natural frequencies, these numbers increased to 40 percent and more than 70 percent, respectively. In terms of frequencies, it is easier to see that there are 10 men with an identical DNA profile in the population, and an additional 100 men where a match is reported although there is no identical DNA profile. That means, only 10 of 110 men for whom a match is reported actually have the DNA profile.

Consider now Question 2, concerning the probability that the defendant is actually the source of the trace found at the crime scene. When the information was in terms of probabilities, a similarly small number of students

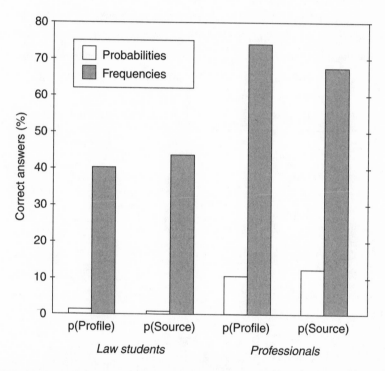

FIGURE 10-2. *How the representation of uncertainties (probabilities versus frequencies) affects the reasoning of legal students and professionals.* When the information about DNA evidence was presented, as usual, in terms of conditional probabilities, few could draw the correct (Bayesian) inference; with natural frequencies more than 40 percent of the students and the majority of the professionals "saw" the answer. Two questions were asked: p(Profile), whether a reported match indicates that the defendant actually has the profile (a true match); and p(Source), whether a reported match indicates that the defendant is the source of the trace. These two questions correspond to the first two steps in the chain of inference shown in Figure 10-1. (Adapted from Hoffrage et al., 2000.)

and professionals were able to figure out what it implied. With frequencies, however, insight emerged at the same rate as with Question 1. Many could now understand that only 1 in 110 men reported to match the trace was actually the source of the trace.

The final decision in a criminal trial is about the defendant's guilt or innocence, and it is cast in yes-no terms rather than in probabilities (Question 3). Did the representation of the DNA evidence make a difference to their verdicts? Yes. More students and professionals voted "guilty" when the evidence was presented in terms of probabilities, that is, when their minds were clouded (Figure 10-3). This effect was slightly larger among the students, but it was observed in both groups. Overall, guilty verdicts increased by 50 percent when the law students and professionals looked at probabilities.

This study illustrates two points: the low level of understanding of the

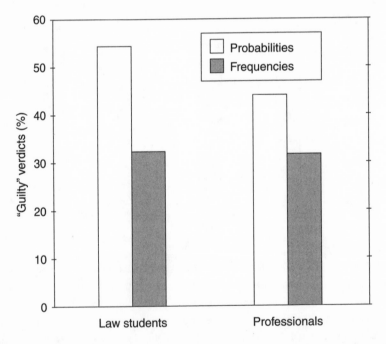

FIGURE 10-3. *Does the representation of uncertainties (probabilities versus frequencies) affect the verdict (guilty or not guilty) of legal students and professionals?* When the information about DNA evidence was presented in conditional probabilities, more students and professionals judged "guilty." (Adapted from Hoffrage et al., 2000.)

uncertainties involved in the chain of inference (Figure 10-1) by law students and professionals, and how an effective tool—natural frequencies—can help overcome innumeracy in the legal profession.[24]

Paternity Uncertainty

Apart from criminal cases, the second major use of DNA testing is in paternity testing. DNA fingerprinting is a significant improvement over the earlier ABO blood group analysis, which could sometimes exclude the possibility that a man was the father, but never come close to proving fatherhood. Before DNA testing, court proceedings were often colored by the public humiliation of unmarried mothers, in which the man who denied being the father and his friends tried to prove that the mother was of bad character. For instance, until the Legitimacy Act of 1959 was passed in England, hearings usually took place in open court, to which the public and the press had a right of admission, and the evening papers provided detailed reports including the names and addresses of the persons involved.[25] DNA testing has helped to dispense with the humiliating character that court hearings had in the past. Now that DNA evidence for paternity exists, courts rarely subject mothers to cross-examinations about their sex lives. The prospect of a DNA test can change denial into confession, as the following case in Great Britain illustrates.

Both parents were aged 18. The putative father's mother tried to remain in court during the initial hearing and had to be removed by the usher. On leaving, she shouted to her son "say what I've told you to say." He denied paternity, and the case was adjourned for him to arrange legal aid and consider DNA tests. He admitted paternity immediately at the next hearing, said he had no wish to see the baby, and was ordered to pay maintenance of £1 a week.[26]

My point is that the mere possibility of DNA fingerprinting can be sufficient to end denial, to spare the mother an inquisition into her sex life, and to oblige the father to pay child support. Yet this British court seems to have assumed that baby care is a rather low-cost proposition.

In order to estimate the probability that a man is the father of a particular child, one needs a prior probability, or base rate, as in medical diagnoses (see Chapter 5). But what could that prior probability be? Many laboratories deal with this problem simply by assuming that the nongenetic evidence in every paternity case indicates that there is a prior probability of .5 that the defendant in a paternity case is the father.[27] The laboratories defend this arbitrary value by citing the *principle of indifference:* either the alleged father is the father or he is not, therefore the prior probability is .5 for each possibility. This practice is controversial because it implies that, in every paternity case, the defendant is as likely to be the father as all other men put together. The principle of indifference has a long history in the law. In the early nineteenth century, jurisprudence was still one of the main arenas for the application of probability theory, such as in the evaluation of the credibility of witnesses. One issue, which then as today divided liberals from conservatives, was how to weigh the two possible errors a judge can commit: to convict an innocent person or to acquit a guilty person. If one tries to minimize the possibility of convicting innocent defendants, one has to pay the price of acquitting more guilty persons as well. If, however, one tries to minimize the possibility of acquitting guilty persons, one will pay the price of imprisoning more innocent ones. The French mathematicians Denis Poisson and Pierre Laplace advocated the conservative view over the earlier liberal reforms envisaged by Condorcet, the Enlightenment philosopher and politician. Poisson emphasized societal security over individual rights even more than Laplace, arguing that the prior probability of a defendant's guilt should be at least .5. In the supposed service of jurisprudence, twentieth-century DNA paternity testing assumed a similar moral stance.[28]

Paternity testing can also be of concern to fathers. A man can never really be sure that he is the biological parent of his children, unlike mothers, who can be sure, except for the possibility of babies being switched in the hospital. As pointed out in Chapter 8, recent DNA analyses of populations in Western countries indicate that some 5 to 10 percent of children have a different genetic father from the man who thinks he is the father. Despite its usually being to the benefit of the children that such deception remain veiled, the market for paternity testing is growing. According to the American Association of Blood Banks, 240,000 paternity tests were performed in the United States in 1997, more than three times as many as 10 years before. Above one of Chicago's busiest motorways, a neon-pink billboard asks "Who's the father?" Call a certain phone number and the answer is available for $500, provided one supplies cheek swabs from mother, child, and the alleged father.[29]

Uncertainties Don't Go Away with DNA

Like every new technology, DNA fingerprinting not only reduces old uncertainties, such as paternity, but poses new uncertainties as well.

WHAT IS THE POPULATION OF SUSPECTS?

In the case of Christina's murder, described at the beginning of this chapter, the police had information from a former victim to generate a hypothesis about the population of which the perpetrator was a member (that is, men between 18 and 30 who lived in Oldenburg and its environs). This is not to say that this hypothesis was certain, as one can infer from the fact that the murderer turned out to be 30 years old—just at the fringe of the age range that defined the suspect population. Yet, it worked in the case of Christina. In other criminal cases the definition of a reasonable population of suspects may be less clear. However, if one wants to compute a source probability—that is, to move from the second to the third link in the chain of inference—one needs a base rate, that is, a population of suspects. (Courts tend not to accept a merely subjective prior probability.) One way to deal with this uncertainty is to make a lower and an upper estimate of the population size, that is, to start with two extreme base rates. For instance, one could consider the 18,000 men in the age range 18 to 30 as the lower base rate, and the larger number of men age 16 to 50 as the upper limit. This would allow the calculation of a lower and upper bound on the source probability.

This problem of specifying the population of suspects exists not only in criminal cases, but also in paternity testing, as we have seen in the previous section. Uncertainty about the population of suspects remains a fundamental issue in making an inference from a DNA match to a source probability.

DOES THE SUSPECT HAVE BROTHERS?

The random match probability is for unrelated people. Therefore, a problem in interpreting a match is the possibility that a defendant has close relatives. For instance, identical twins have the same DNA profile, and close relatives have more similar DNA profiles than unrelated individuals. In one case in Scotland, a forensic scientist reported a DNA match probability of 1 in 49,000 for unrelated individuals, but a match probability of 1 in 16 for a brother of the defendant. It happened that the defendant had five brothers.[30] Thus, if there is no further evidence that distinguishes among the brothers, the DNA match per se is only weak evidence. Close relatives who could have committed a crime or parented a child need to be figured into the interpretation of a DNA match.

HOW CERTAIN IS A MATCH?
A RANDOM MATCH PROBABILITY?

A further reason for caution about the DNA profiling process is that subjective judgments are an integral part of it. These arise both in the judgment as to whether or not there is a match and about the size of the match probability. As mentioned earlier, a match is declared when the bands of two DNA profiles line up with one another. However, because they do not always line up exactly, a match is defined when two profiles line up within a specified error range. Thus, a match is not a black-and-white observation, but depends on what deviation one tolerates. This arbitrary cutoff point has the undesirable consequence that one profile deemed not to match can be arbitrarily close to another that is declared a match.

Similarly, the random match probability is not just a matter of looking up the population statistics. For instance, in 1987, Vilma Ponce and her 2-year-old daughter were stabbed. José Castro, a Hispanic male and neighbor, was interrogated, and a small bloodstain from his watch was sent for analysis. Castro claimed that the blood was his own. Lifecodes Corporation, which carried out the test, however, reported a match between the DNA in the bloodstain and the victim's blood, and reported a random match probability of smaller than 1 in 100 million. Harvard University and the Massachusetts Institute of Technology examined the same data and arrived at a probability of 1 in 24.[31] Such huge discrepancies can arise when samples are corrupted by chemical or biological contaminants, or exposed to potentially harmful environmental factors such as sunlight, and when laboratory practices are poor or results inadequately analyzed.

A THREAT TO PRIVACY?

The biological function of DNA molecules is to provide a genetic message, the gene, encoded in a sequence of bases. Therefore, DNA databases contain information of such a personal nature that they seem to pose a significant threat to privacy. Some people fear that if we learn to "read" DNA, then others will be able to read our genetic future like a script. To judge the reality of this fear, we need to keep in mind that DNA databases do not contain entire DNA sequences, only DNA profiles, which contain information about several loci (sites) rather than the entire genome. Moreover, the databases generally contain "noncoding" loci, that is, loci that are not believed to contain genes. Recall that part of the DNA—sometimes referred to as "junk DNA"—does not seem to have coding functions as genes have. The reason for focusing on noncoding loci is not to protect privacy. Instead, it is to discriminate better between individuals.[32] Thus, given current technologies, DNA databanks afford little information about any particular individual.

That said, we need to be cautious about predicting that the future will be like the present. Historically, new technologies tend to change conceptions of civil liberty. When ordinary fingerprinting was first introduced in the United States, many courts saw its use in jurisprudence as a violation of civil rights. It is not inconceivable that, in the near future, mouth swabbing of job applicants, health insurance applicants, and immigrants will be considered as unproblematic as fingerprinting is now. Unlike the human genome project, which catalogues the coding part of DNA, data banks suitable for DNA fingerprinting contain samples of noncoding DNA.

There are obvious benefits from the latter for quickly identifying the perpetrator in sexual assault cases and other criminal cases. The availability of the technology supported by data banks might, in the long run, deter some potential criminals, at least those who know they are recorded in the data bank. Even if there were only little or no deterrence, data banks will help to determine criminals after the crime and keep them from committing multiple crimes.

New Technology Can Alter the Law

The traces on O. J. Simpson's gloves and Monica Lewinsky's blue dress[33] have made Americans familiar with DNA evidence. At the same time, the introduction of DNA evidence in criminal cases has put pressure on the law and new demands on legal professionals.

The statute of limitations for rape prosecutions is one example of pressure for change. For instance, in New York, rape cases expire after five years. That is, there is no possibility of prosecuting a case after it is more than five years old. As one victim, a Lower East Side student who was raped two years ago and whose case is unsolved put it: "I don't understand how there can be a timetable on this crime. Rape has completely changed my whole life."[34] The rationale for this statute of limitations was to give defendants a fair chance because as time goes by, evidence disappears and memories fade, and consequently, defendants will face difficulties proving where they were on a particular day or at a particular hour many years before. This historic rationale is now being strained by the availability of DNA fingerprinting, which can give much more conclusive evidence than any technology before. As a consequence, several states, including Florida, Nevada, and New Jersey, have recently abolished the time limits in cases of sexual assault. Science has revolutionized the investigation of rape cases, and the law is catching up with the new situation.

A second law that may be put under pressure through the introduction of DNA fingerprinting is the death penalty. Among the large democracies, only the United States, India, and Japan still put prisoners to death. A Gallup poll in 2000 indicated that the proportion of Americans favoring the death penalty was decreasing (although some two-thirds are still in favor if it) and the number of people who believe that innocent people are at least occasionally wrongly sentenced to death is increasing. The Gallup researchers suggest that this changing situation is linked to the advent of DNA technology, which in some cases has shown that innocent people are being sentenced to death.

DNA fingerprinting has also put new demands on the legal profession. These include overcoming the illusion of certainty and learning how to understand and communicate uncertainties. In this chapter, I have pointed out various sources of uncertainty in the chain of inference from a DNA match to a guilty verdict, and showed that the same tools that helped physicians and counselors can help legal professionals. For instance, using natural frequencies instead of probabilities can greatly reduce sources of confusion, such as the prosecutor's fallacy. Our experiments indicate that proper representation of uncertainties not only fosters insight in legal students and professionals, but also changes the number of "guilty" verdicts they are willing to deliver.

Now it is up to the courts to recognize the relevance of how to represent uncertainty in trials. A few attorneys and judges in Great Britain and America have taken notice. A British Court of Appeal recently recommended that DNA evidence be presented in a frequency format.[35] However, this recognition is still occurring in an accidental way rather than as a systematic part of legal training and procedures. For instance, one of my colleagues who had been writing an article for a legal journal on the importance of frequency representations happened to be acquainted with an attorney who was then on the O. J. Simpson defense team. As a consequence, the defense team asked Judge Ito not to allow the prosecution DNA expert, Professor Bruce Weir, to testify in terms of conditional probabilities and likelihood ratios, which are ratios of conditional probabilities. The team requested that the jurors hear the information in frequencies instead, arguing that they might confuse random match probabilities with statements about the probability that Mr. Simpson was actually the source of the samples.[36] Judge Ito and the prosecution agreed. The prosecution expert then used likelihood ratios anyway! Statisticians (such as Weir) are not always aware of the psychology of risk communication.

Can we know the risks we face, now or in the future?

No, we cannot; but yes, we must act as if we do.

Mary Douglas and Aaron Wildavsky,
Risk and Culture

11

VIOLENT PEOPLE

In Chapter 1 we met a psychiatrist who told his patients that they have a 30 to 50 percent chance of developing a sexual problem when taking Prozac. The psychiatrist and his patients understood this statement differently. For the psychiatrist, it referred to the proportion of his patients who develop such a problem as a result of taking the drug. For the patients, it referred to the proportion of their sexual encounters in which something would go wrong. Each side construed the reference class from its own point of view—for the psychiatrist, the reference class was "my patients," whereas for the patients, it was "my sexual encounters." A probability of a single event leaves open what the *reference class* is—by definition. But people tend to assume, at least implicitly, a specific reference class when they try to understand a single-event probability, opening the door to interpretations that contradict one another and the intended meaning. The same ambiguity can occur in legal and clinical settings in which the aim is to predict violent behavior.

Foretelling Violence

Violence is part of our lives. People are afraid to walk in their neighborhoods at night. Kids with guns wound and kill their peers and teachers in

American schools. Adolescents and young right-wingers attack foreigners in the former East Germany with knives, fists, and boots. Wife battering and the beating of children are part of family life in modern Western societies. Why and when are people turning to violent acts?

Forecasting violent behavior is difficult. According to the American Psychiatric Association in a communication to the U.S. Supreme Court, its "best estimate is that two out of three predictions of long-term future violence made by psychiatrists are wrong."[1] Nevertheless, the U.S. Supreme Court has ruled repeatedly that such testimony is legally admissible as evidence. The Court concluded that mental health professionals' predictions were "not always wrong . . . only most of the time."[2] The weakness of these forecasts illustrates a larger problem. When expertise plays a role in criminal justice, it is often in situations that are badly understood and where testimony is only marginally grounded in scientific knowledge.

Probation officers are professionals who supervise criminal offenders who have been granted conditional freedom, whether they are on parole, bail, probation, or weekend release. A probation officer faces the task of foretelling violence. He or she can make either of two errors: recommend freedom for someone who will commit another violent act or deny freedom to someone who will not commit such an act. An error of the first kind was committed by the expert who recommended releasing the man who eventually murdered Christina, as related in the previous chapter.

How do experts such as probation officers advise the court about the potential risks posed by a criminal offender? Until recently, experts have used yes-no labels such as "dangerous" and "not dangerous." In a Texas capital sentencing case, for instance, Thomas Barefoot was convicted of killing a police officer after being stopped in an arson investigation. One of the experts who testified at Barefoot's sentencing hearing, a psychiatrist popularly known as Dr. Death, commented that it was "one hundred percent and absolute[ly]" certain that Barefoot would be violent in the future.[3] (The psychiatrist had neither interviewed nor even met Barefoot.) As in weather forecasting, however, statements that acknowledge uncertainty have since become more common in professional predictions concerning violent behavior. For instance, the psychologists John Monahan and Paul Slovic and others have advocated that mental health professionals use probability estimates, rather than verbal labels, to communicate the chances that violent behavior will occur.[4] As a result, probation officers were asked to estimate the probability that a person would commit another violent act if he were given parole or probation.

Yet these experts were reluctant to use numerical probabilities to communicate risk of violent behavior. They preferred categorical labels, believing that there was too much uncertainty to justify numbers and did not want to be held accountable for being precise. In a series of seminal studies, Slovic and Monahan investigated experts' numerical risk judgments, with striking results.

SINGLE-EVENT PROBABILITY VERSUS FREQUENCY JUDGMENTS

In one study, Slovic, Monahan, and MacGregor tested more than 400 members of the American Academy of Psychiatry and Law (AAPL) and more than 400 members of Division 41 of the American Psychological Association, that is, the American Psychology-Law Society. These experts received information about four actual patients in the form of one-page abstracted discharge summaries taken from the records of patients discharged in 1996 from an acute care inpatient facility.[5]

What follows is the first half of one discharge summary for a woman described as restless, anxious, and having poor judgment and concentration.

Case 22-190 Discharge Summary

History of Present Illness

This is a 52-year-old divorced Hispanic female with a history of psychiatric illness, including bipolar and schizo-affective disorders. She has a low level of cognitive functioning, and is borderline mentally retarded. Her sister states that the patient struck her several times on the day of admission because the sister opened the window to let some smoke out. The sister had to grab onto the patient's shoulder and bring her to her room. The patient complained, "My sister grabbed my face and hurt me." The patient has been paranoid of her sister, and has had auditory hallucinations. She has also had illusions of a man at her window and sitting in her living room, as well as illusions of lice in her hair. She has been scratching and picking at her head, and washing herself excessively. The patient denies [feeling homicidal and suicidal], and has been compliant with medications.

Family and Social History

The patient is divorced with four children. Her ex-husband is an alcoholic. She currently lives in her sister's house with her sister's husband, sister's father, and the patient's brother. The brother carries a diagnosis of schizophrenia. The sister also cares for a grandson who has severe mental retardation. The patient lived in an orphanage as a child, and has a history of being sexually abused. She was taken out of school in the first grade because of an inability to perform.

The second half of the discharge summary describes the medication the patient is receiving and her family's wish to have her return home. Is this woman likely to harm someone other than herself during the six months following discharge?

One half of the experts were asked to judge the probability that this woman (or one of the other three cases) would harm someone other than herself during the six months following discharge, that is, would commit a violent act. The other half were asked to judge how many out of 100 women like this woman would commit a violent act in the same period. One might think that the answer would be about the same because all professionals read the same discharge summary. This was not the case. The two questions produced two systematically different responses. The probability judgments were about 50 percent higher than the frequency judgments. Figure 11-1 shows the predictions of violent acts averaged across the four patients. The average frequency estimate of the members of the American Academy of Psychiatry and Law was 20 out of 100, whereas their average probability estimate of a violent act was .30. The members of the American Psychology-Law Society estimated the risk of violent acts generally higher, but showed about the same difference between judgments of frequencies and probabilities.

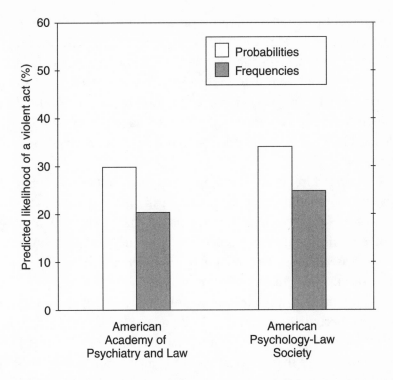

FIGURE 11-1. *The predicted average likelihood that a patient will commit a violent act.* More than 400 members of the American Academy of Psychiatry and Law (left) and more than 400 members of Division 41 of the American Psychological Association, that is, the American Psychology-Law Society (right), predicted the likelihood that each of four patients would commit a violent act. One half of the members in each group were asked for a probability judgment, the other half for a frequency judgment. The members of the American Psychology-Law Society had, in addition, received a tutorial on the definition of harm and probability theory. Nevertheless, both groups of experts made systematically different predictions when asked for frequency and probability judgments. (After Slovic, Monahan, and MacGregor, 2000.)

It is hard to believe that professionals' estimates of the chances of a patient harming someone would rise when they are asked to give a probability. Is there an explanation? Assume that these professionals, like many ordinary people, think of probabilities in terms of classes of concrete cases. As in the Prozac case in Chapter 1, one's interpretation of a single-event probability depends on the reference class to which one implicitly assumes the probability refers. One may take the reference class to be a set of events, such as the set of repeated occasions on which a particular patient leaves the hospital on weekend release and then returns. In that case, the question will be interpreted as being about one patient who is repeatedly granted conditional freedom. Alternatively, one may take the reference class to be a set of patients who are granted weekend release on one occasion. The answers to the question "What is the probability that this person will commit a violent act in the next six months" need not be the same given the two interpretations.[6]

In contrast, the frequency question "How many out of 100 women like this patient would commit a violent act in the next six months?" clearly specifies one reference class: other patients like the violent patient. This argument explains why differences between frequency and probability judgments can arise, but not the direction of the difference. With respect to the question of direction, that is, why probability judgments are numerically higher than frequency judgments, I can only offer speculation, not an answer. If an expert assumes that the likelihood of a violent act increases the more often a patient is granted conditional freedom, such as weekend release, then it follows that the relative frequency of violent acts of 100 patients given conditional freedom one time (that is, the first time) should be lower than that of one patient given conditional freedom multiple times. Because the probability question leaves open which of the two situations is meant, this speculation can explain why frequency judgments resulted in lower predicted risks.

Judgments of single-event probabilities and frequencies can differ systematically—in the case of the former, the reference class can be interpreted in many ways; in the case of the latter it is clearly specified. The ambiguity introduced by a question such as "What is the probability that this person will commit another violent act?" can lead to systematically different judgments on the part of experts, as was found in the study summarized above.

DO RESPONSE CATEGORIES INFLUENCE EXPERTS' JUDGMENTS?

Probabilities of violent behavior, like those of rain, are too imprecise to be expressed by numbers like 31 percent. They are more readily expressed in categories, such as 5 percent, 10 percent, 20 percent, and 30 percent. Assume you are designing a form showing a set of categories that experts can use to judge how dangerous a patient is. Which categories should you use? Is this choice merely aesthetic, or would the spacing of these categories matter to the resulting judgments? To answer this question, Slovic and his collaborators gave each member of the American Academy of Psychiatry and Law who participated in their study one of the two response scales shown in Figure 11-2. Members in one group made their predictions of violent acts using response categories of 1 percent, 10 percent, 20 percent, and so on in increments of 10 percentage points ("Large probabilities"); members in the other group made theirs with categories that were more finely graded at the lower end ("Small probabilities").

Does the spacing of categories influence judgments of violent acts? In an earlier study, Slovic and Monahan had found that probabilities of harm were indeed systematically higher when people were given the scale with large probability categories (Figure 11-2).[7] I will call this effect of the spacing of category on judgment the "category effect." However, this category effect was reported for laypeople making predictions of harm for hypothetical persons described in vignettes, not experts considering real cases. Would the choice of categories also influence professionals' judgments of real cases?

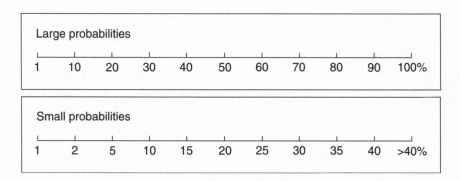

FIGURE 11-2. *What is the probability that the patient will harm someone?* Two response scales showed different categories for probabilities. (From Slovic, Monahan, and MacGregor, 2000.)

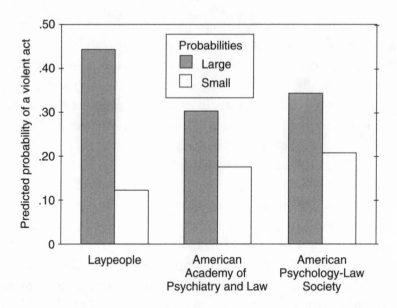

FIGURE 11-3. *Does the choice of response scale influence judgments of the predicted proba-bility of a violent act?* Although there are differences between laypeople and experts, the choice of response scales (large versus small probabilities, see Figure 11-2) does consistently influence judgments. The members of the American Psychology-Law Society had in addition received a tutorial on the definition of harm and probability theory. (After Slovic et al., 2000.)

As shown in Figure 11-3, the members of the American Academy of Psychiatry and Law exhibited the same type of category effect found in laypeople. In experts, the effect was smaller but still substantial. One might suspect that the experts did not really understand what a probability judgment is. To test for this possibility, Slovic and his colleagues gave the members of the American Psychology-Law Society a tutorial in probability theory that explained what is meant by harm, what a probability is, and how probabilities are assessed. It even included an explanation of and a warning about the category effect. Yet this tutorial made no difference: The category effect was about the same size among the experts who had received the tutorial as among those who had not. When the participants were further divided into those who frequently make numerical risk assessments in their profession and those who never or rarely do so, the category effect remained equally strong in both groups.[8]

Thus, the choice of response categories influences the probability judgments of experts as well as laypeople. A similar category effect was found when experts were asked for judgments of frequency. However, in neither experts nor laypeople did the category effect influence the ordering of patients according to judged harm. These results indicate that experts can reliably judge whether a patient is more dangerous than another, but their quantitative estimates are influenced by the categories with which they express those estimates. In other words, the experts can reliably give judgments about the order of patients with respect to the probability of their doing harm but not the *magnitude* of the probability of harm.

Reliability does not imply accuracy. That is, the fact that a particular ordering of patients with respect to probability of doing harm remains stable across the two category scales does not imply that this ordering reflects the actual ordering of the patients with respect to whether they later commit violent acts. Slovic, Monahan, and MacGregor did not report how accurate the experts' judgments proved to be. However, recall the American Psychiatric Association's estimate that two out of three predictions are wrong; furthermore, studies of actual parole decisions suggest that the predictive accuracy of these judgments is quite low.[9]

RESPONSE CATEGORIES AS CUES

The category effect is not limited to forecasting violence. It has been observed in judgments concerning events other than risks, such as in reports about innocuous habits and behaviors.[10] The effect seems to occur in situations that involve uncertainty, such as when a person has limited knowledge and is asked to report or predict behavior. A category effect would not occur in situations where people have, or believe they have, definite knowledge, such as when a person is asked "How many kids do you have?" Here, the person would simply scan the scale in search of the right number. Situations that involve some degree of uncertainty, however, are abundant.

One such situation is survey studies, in which respondents are asked to report their behavior by selecting a category. For instance, Norbert

Schwarz and his colleagues asked German adults how many hours a day they spend watching TV on a typical day.[11] The researchers gave one group of participants a response scale with six categories "up to ½ hour; ½ to 1 hour; 1 to 1½ hours; . . . more than 2½ hours." This was called the "low-category" scale. A second group was given a high-category scale on which the six categories were "up to 2½ hours; 2½ to 3 hours; . . . more than 4 hours." The choice of categories influenced people's reports of how much time they spent watching TV. For instance, 16 percent of the participants given the low-category scale said they watched more than 2½ hours per day. Among those given the high-category scale, 38 percent reported doing so. This is a remarkable phenomenon, and it once more demonstrates how judgments are influenced by the choice of the category scale. Similarly, the reported frequency of sexual behaviors, of consumer behaviors, and of physical symptoms is influenced by the categories given to the respondent.[12] These results highlight the dangers of taking survey data at face value—even when people are not lying or trying to give answers that reflect favorably on themselves. The category effect is a quite general phenomenon that affects judgments of risk and more.

What is the explanation for the category effect? A necessary condition seems to be that people be uncertain about what the answer to the question really is. People do not keep records of the number of hours they watch TV. When asked to estimate this number given their incomplete knowledge, they can use the response categories as a clue to guide them. Someone who thinks herself an average TV watcher will tend to choose a category in the middle of the range, whatever that range may be. Similarly, a psychiatrist who is uncertain about how the risk of violent behavior is distributed in a population of patients will tend to distribute his judgments over the response categories, assuming that the categories reflect the true range of the behavior in the population. The professional, just as the TV watcher does, acts on the assumption that the researcher knows the range in the general population and has chosen the categories appropriately. Using their social intelligence, they both make the seemingly reasonable inference that the researcher's choice of response categories is relevant to what they have been asked to do. Survey researchers or public relations agencies, however, who want to produce results that are favorable to their economic, social, or political position can easily exploit the category effect and produce results in the direction they would like to see. The measurement instrument is not a neutral mirror of an underlying belief; instead, it becomes part of the judgment.

How can the category effect be reduced in assessments of risk? There are two ways. The first is to reduce the uncertainty in the minds of risk assessors, for instance, by providing them with statistical information about the actual violent behavior of inmates on parole or patients on weekend release. As their knowledge increases, their uncertainty diminishes, and the category effect will eventually disappear. A second way is to dispense with response categories and use other tools for risk assessment, for example, an open-ended response format: "Think of 100 patients like Mr. Jones. How many will turn violent within six months? _____ out of 100." In general, laypeople and experts need to take the category effect into account when interpreting forecasts and survey responses.

CAN A SINGLE-EVENT PROBABILITY BE WRONG?

D. A. Cloud, a 26-year-old Seattle man, was convicted of the shooting death of his former middle-school teacher, who had abused him and other students. At his trial, the prosecution offered him the chance to plead guilty to second-degree murder in exchange for a recommended sentence of 15 years (a plea bargain) rather than stand trial on first-degree murder charges with an uncertain outcome. But according to Cloud, his lawyer—one of Seattle's best-known criminal defense attorneys—told him that he had a 95 percent chance of acquittal of the more serious charge based on an insanity plea.[13] With this high probability in mind, Cloud rejected the offer, stood trial, and was sentenced to 20 years in prison for first-degree murder. The disappointed Cloud filed a motion to set aside this verdict, accusing his lawyer of having given him unrealistic probabilities. After an extraordinary 12-day hearing, the court denied the motion.

Note that the lawyer used a single-event probability to express Cloud's situation: a 95 percent chance of acquittal, which implies a 5 percent chance of conviction. Can such a probability ever be wrong? Can it be unrealistic? It does not seem that it can be wrong because it refers to one case only, that of Mr. Cloud. The defendant will either be acquitted or he will not, and the probability estimate allows for both—unless the probability is stated to be 0 or 1, which it was not in Mr. Cloud's case. A frequency prediction, in contrast, can be wrong. The statement that 95 out of 100 defendants (in similar cases of homicide after abuse) will be acquitted, for example, can be shown to be right or wrong in light of data from legal records.

The Question Is Part of the Answer

Surveys have been designed to measure people's perceptions of everything from their risk of acquiring mad cow disease after having eaten British beef to that of developing cancer from living close to a nuclear power plant. Typically, such surveys use response categories to facilitate the analysis of data, and the underlying assumption is that people have subjective risks in their minds and straightforwardly map them onto the response categories. The response scale is seen as a neutral measurement tool, like a typewriter, which is merely a device for recording thoughts in the form of letters and punctuation marks. The studies reported in this chapter show that this assumption is incorrect. When experts estimate the risk that a violent person will do harm to another person in the near future, the spacing of the categories substantially influences their judgments. So does the choice of question—that is, whether it pertains to a probability or a frequency. Indeed, the question and the response categories are *part* of the answer.

What can be done with fragile probabilities, that is, probability estimates that rest on assumptions about reference classes and response categories? The effect of response categories can be eliminated by using open-ended response formats. Concerning the ambiguity of the reference class, my recommendation is this: *If you want to make a single-event statement, make a frequency statement first.* The frequency statement forces you to clarify what the reference class is and reduces the possibility of miscommunication: "Out of every 100 inmates similar to Mr. Bold, 20 are expected to behave violently within six months if let out on parole. In other words, Mr. Bold has a 20 percent likelihood of violence." The initial frequency statement clarifies the reference class behind the probability statement that follows.

Asking for a probability versus a frequency judgment and providing category spacing can both be used to influence predictions and estimates obtained from laypeople and experts. In the next chapter, I will show how various representations of benefits and costs, such as relative risk reduction, can be used to influence policy decisions by exploiting others' innumeracy. The power of playing with representations in public policy affairs is of deep concern, because it illustrates how people can be exploited without their noticing it. At the same time, however, some of these stories also reveal a funny side, as instances of innumeracy often do when they are, indeed, noticed.

PART III

FROM INNUMERACY TO INSIGHT

If an unfriendly foreign power had attempted to impose on
America the mediocre educational performance that exists
today, we might well have viewed it as an act of war.

The National Commission on Excellence in Education

12

HOW INNUMERACY CAN BE EXPLOITED

In front of me is a wonderful little book entitled *How to Lie with Statistics.*[1]
It begins with five epigraphs, the first of which reads "There are three kinds
of lies: lies, damned lies, and statistics. —Disraeli." Everyone loves this quotation, which was tentatively attributed by Mark Twain to Benjamin Disraeli, a prime minister under Queen Victoria.[2] The second epigraph reads
"Statistical thinking will one day be as necessary for efficient citizenship as
the ability to read and write. —H. G. Wells." Statisticians love this quotation, which has been reproduced in statistical textbooks again and again.
The author of *How to Lie* provides no source for the second quotation—he
simply attributes it to the science fiction writer H. G. Wells, as do the many
others who love to quote it. When I tried to find this quotation in Wells's
writings, however, I couldn't. In any case, *How to Lie* is not actually about
lying, but about how to represent information correctly yet misleadingly.
Similarly, the chapter you have begun to read is not about lying but instead
about choosing representations that mislead the innumerate without being inaccurate. Given that the number of "innumerates" is supposedly le-
gion, the opportunities to mislead them are endless. When John Q. Public
does not understand relative risks, single-event probabilities, or conditional probabilities, it's his own fault, isn't it?

Why doesn't everyone communicate risks transparently? I have asked this question of myself many times. One answer is that innumerates can be exploited. It can be to the advantage of "numerates" to be nontransparent as long the number of innumerates is sufficiently large. As a consequence, transparent risk communication is unlikely to emerge as long as a large proportion of people are innumerate. How do the numerate exploit others' innumeracy?

How to Get Funding

Why do researchers and health agencies continue to report the benefits of treatments in terms of relative risks, even though this type of risk communication is known to mislead people about the degree of a treatment's benefit?

A team of British researchers studied the decision making of members of the Anglia and Oxford regional health authorities. This group included executive members, who are responsible for purchasing and financial and personnel management, as well as nonexecutive members, who are appointed by the Home Secretary to act as public representatives of their local populations. The members were asked whether they would support purchasing each of four cardiac rehabilitation programs and each of four breast cancer screening programs.[3] The alternative programs had the same benefits and differed only in the way the benefits were described. For each program, one proposal reported relative risk reduction, the second absolute risk reduction, the third the number of patients who need to be treated to prevent one death, and the fourth the proportion of "event-free" (that is, surviving) patients.

Table 12-1 shows data about the relative effectiveness of coronary artery bypass surgery (the cardiac rehabilitation program).

Four Ways to Present the Benefit

The absolute risk reduction of bypass surgery is 4.1 percent (404 − 350 = 54; 54/1,325 = 4.1%).

The relative risk reduction of bypass surgery is 13.4 percent (4.1/ 30.5 = 13.4%).

The percentages of event-free (surviving) patients are 73.6 percent and 69.5 percent, respectively, for surgery and medical therapy.

The number of patients needed to be treated to save one life is 25.

Whereas the relative risk reduction with bypass surgery is 13.4 percent, the absolute risk reduction is 4.1 percent. The number of patients who need to be treated with bypass surgery to save one more life than by medical treatment is 25 (note that 1 out of 25 corresponds to 4.1 percent). In other words, out of each 25 patients who receive a bypass, 1 patient's death (within 10 years) will be prevented; the other 24 will have no benefit in terms of mortality reduction from the operation. Finally, the number of event-free patients (here the number of patients who survive) is 73.6 percent for bypass surgery versus 69.5 percent for medical treatment. All these are equivalent descriptions of one and the same outcome of an actual randomized trial in which the two treatments were compared.

One might think that in deciding which treatment programs to fund, the British health authorities would not have been influenced by differences in representations of the same outcome. As it turned out, though, they were. In the case of both cardiac rehabilitation and breast cancer screening, the authorities saw the program as having the greatest merit—and were most willing to fund it—when its benefits were expressed in

---

TABLE 12-1: *What are the benefits of coronary artery bypass surgery versus medical therapy?* This table shows the actual results of a clinical study; the text shows four ways to present the result. (After Fahey et al., 1995.)

Treatment	No. of Patients	Deaths
Coronary artery bypass surgery	1,325	350 (26.4%)
Medical therapy (nonsurgical)	1,324	404 (30.5%)

terms of relative risk reduction (Figure 12-1). When its benefits were expressed in terms of absolute risk reduction or the proportion of event-free patients, the authorities saw the treatments as having the least merit. (After all, 13.4 percent is more than 4.1 percent.) When the benefits were reported in terms of the number of patients who needed to be treated to save one life, their willingness to fund was in-between. Given these results, one might wonder how many of these professionals noticed that the benefits of the four programs were the same. In fact, only 3 out of the 140 experts realized that all four reflected the same clinical results.

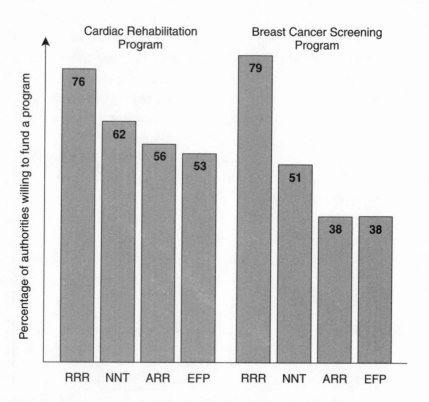

FIGURE 12-1. *Does the willingness of members of health authorities to fund a program depend on how the benefit is presented?* Willingness to fund is highest when the proposal reports benefits in terms of relative risk reduction (RRR), followed by number needed to treat (NNT), absolute risk reduction (ARR), and event-free patients (EFP). (After Fahey et al., 1995.)

How to Sell Your Treatment

In Chapter 5 we learned of a study of the information brochures on breast cancer screening distributed by Australian health organizations in which it was found that the benefits of screening, when they were reported at all, were expressed in terms of relative risks. These leaflets were written for the public, not for professionals. In no case were the benefits of screening explained so that laypeople could easily understand them. Similarly, the benefits of a diet or medical treatment are often reported in terms of a relative risk reduction or relative risk increase. Consider a newspaper article in which it is reported that men with high cholesterol have a 50 percent higher risk of heart attack. The figure of 50 percent sounds frightening, but what does it mean? It means that out of 100 fifty-year-old men without high cholesterol, about 4 are expected to have a heart attack within 10 years, whereas among men with high cholesterol this number is 6. The increase from 4 to 6 is the relative risk increase, that is, 50 percent. However, if one instead compares the numbers of men in the two groups who are *not* expected to have a heart attack in the next 10 years, the same increase in risk is from 96 to 94, that is, about 2 percent. Now the benefit of reducing one's cholesterol level no longer looks so great. This trick doesn't work with absolute risks. The absolute risk increase is 2 out of 100, or 2 percent, no matter whether one counts those with or without heart attacks. Absolute risks do not leave room for playing with such numbers.

Consider a woman trying to decide whether or not she should have hormone replacement therapy. Hormone therapy has potential benefits and costs. If one wants to help patients to make an informed decision, these should be reported in the same "currency," such as absolute risks. But some physicians would like to influence women's decisions, not only inform them. One technique is to exploit patients' ignorance concerning risk representations. To increase the patient's willingness to participate in therapy, one simply reports the benefits in terms of relative risks (which appear larger) and the costs in terms of absolute risks (which appear smaller). To decrease the patient's willingness, one simply does the opposite. To take an example, the following information leaflet written by 12 physicians was made available in the waiting rooms of German gynecologists.[4] The relevant passages are translated below.

Hormones and Cancer
Up-to-Date Information

Dear Patient,

The media continually report a threatening increase in cancer in connection with the use of hormone replacement therapy during menopause. In what follows we give you an up-to-date review of the proven facts so that you have an objective basis for making a decision.

Breast Cancer. About 60 extensive studies on this topic have been conducted to date. The results are not unanimous. A summary of these studies shows that hormone therapy may be associated with a minimal increase in the incidence of breast cancer.

Usually, about 60 out of 1,000 women develop breast cancer in their lifetime; after 10 years of hormone therapy, 6 more women develop breast cancer. That is, the risk may possibly increase by 0.6 percent (6 in 1,000). . . .

Other Cancers. Not only does hormone therapy not increase colorectal cancer, which is relatively frequent, but it has been proven to protect women against colorectal cancer (by up to more than 50 percent). That is, women who receive hormone therapy develop colorectal cancer only half as often. . . .

It is easy to guess in which direction these physicians were trying to influence their patients. The potential cost (an increased risk of breast cancer) was reported as an absolute risk and the potential benefit (a decreased risk of colorectal cancer) as a relative risk, making the cost appear smaller and the benefit larger. If these physicians had reported both the benefit and cost in relative risks, the "minimal increase" in breast cancer would have been 10 percent, not 0.6 percent. In addition, the cost was marked with the qualifier "possibly," whereas the benefit was described as "proven." This choice of words and representations is unlikely to have been accidental. Rather, these physicians were almost certainly trying to exploit their patients' inability to see that apples and oranges were being compared—with the aim of getting them to agree to hormone replacement therapy.

How to Raise Anxiety

Since the introduction of the contraceptive pill in the 1960s, women have gone through several "Pill scares." A few years ago, information concerning side effects of oral contraceptives was publicized in Britain. The official statement said that "combined oral contraceptives containing desogestrel and gestodene are associated with around a two-fold increase in the risk of thromboembolism."[5] ("Thromboembolism" means blockage of a blood vessel by a clot.) The warning, which was phrased as a relative risk increase, caused great concern among women and their physicians. Many women stopped taking the Pill, which resulted in an increase in unwanted pregnancies and abortions.[6]

If the same information about thromboembolism had been expressed as an absolute risk, it would have been clear how frequent this dangerous side effect actually is. The relative risk says only how much more likely thromboembolism is if one takes the Pill than if one does not take the Pill, not how often thromboembolism actually occurs. In terms of absolute risks, the chance of thromboembolism increases from about 1 to 2 in 14,000 women.[7] In terms of relative risks, the chance doubles.

Reporting relative risks can raise people's anxieties. These anxieties, in turn, can change people's behavior and encourage them to disregard the side effects of not taking the Pill, such as abortions and unwanted pregnancies, that also pose hazards to their health. Note that the choice is not between risk (thromboembolism) and certainty (no thromboembolism), but between two options, each with its own set of risky consequences. Absolute risks help women understand how often each of these consequences may occur. Once again, however, transparency helps reduce unnecessary anxieties, and abortions as well.

How to Make More Money

Assume you suffer from a serious illness and that without proper medication, you will surely die. The medication you have been taking so far reduces your risk of dying to a probability of .0006, at a cost of $185. The company that produces this medication plans to market an improved version that would reduce your risk to .0003. How much would you be willing to pay for the improved version?

This question was posed to a group of Swiss university students. They said they were willing to pay a slightly higher price (an average of $213) for the improved medication. Another group was given the risk reduction in absolute frequencies, that is, the group was told that the new medication would reduce the risk of dying from 600 to 300 out of 1 million. These students were willing to pay substantially more (an average of $362) for the improved medication.[8] Thus, when the benefit was represented in frequencies rather than single-event probabilities, the perceived monetary worth of the improved medication was substantially higher.

Two caveats should be mentioned here. First, the possibility of increasing the perceived worth of a product by expressing risk reduction in frequencies rather than single-event probabilities is not boundless, especially when the hazard is extremely rare in the first place. For instance, the students' willingness to pay more when the risks were explained in frequencies disappeared when the risks were extremely small (for example, 3 or 6 in 1 million). Second, unlike most studies covered in this book, this study tested students rather than experts and hypothetical illnesses rather than real ones. As a consequence, the participants' statements about their willingness to pay had no real consequences for them. It remains to be shown whether one can use this technique to increase profits from real patients with real illnesses who need real medications. But expressing risk reduction in frequencies may enable businesses to charge consumers more for health-enhancing products.

How to Present Losses as Gains

Your firm has had quite a few ups and downs in the last three business quarters. Your eager-beaver assistant has made a plot of sales over this period (Figure 12-2) that shows that your firm experienced an overall loss in sales. Aside from your assistant, you are the only person who has seen this graph. It is in your best interest not to make this loss transparent because you fear it will make some of your stockholders nervous. What should you do?

Instead of publishing the graph, you describe it verbally. You release a press notice stating that between January and May, your firm suffered a 50

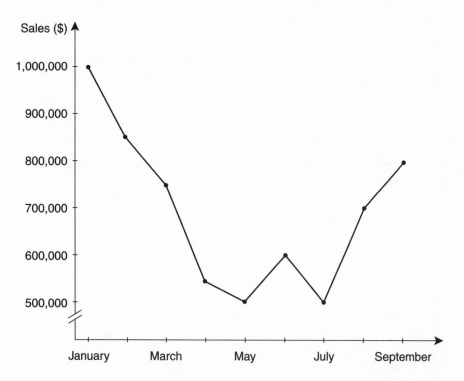

FIGURE 12-2. *Losses and gains.* The loss in sales visible in this figure can be masked by the statement "There was a 50 percent loss in the first part of this period, but this was compensated for by a 60 percent increase in the second part." However, a 50 percent loss from January to May is not cancelled by a 60 percent gain from May to September.

percent decrease in sales, which was regrettable but consistent with the market as a whole. However, the notice says, by taking countermeasures, the firm was able to compensate for this loss. Between May and September, the firm increased its sales by 60 percent. Overall, then, the dip in sales in the first five months was more than outweighed by a steeper climb in the following months. Thus, according to the notice, the overall result looks positive because 60 percent is more than 50 percent. In actuality, the initial 50 percent loss was $500,000, and the later increase of 60 percent amounted to only $300,000.

In the late 1970s, the Mexican government faced the problem of how to increase the capacity of the Viaducto, a four-lane motorway.[9] Rather than building a new highway or extending the existing one, the government implemented a clever, inexpensive solution: It had the lines on the four-lane highway repainted to make it six-lanes wide. Increasing the number of lanes from four to six meant a 50 percent increase in capacity. Unfortunately, the much narrower lanes also resulted in an increase in traffic fatalities, which, after a year, forced the government to turn the highway back into a four-lane road. Reducing the number of lanes from six to four meant a 33 percent decrease in capacity. In an effort at touting its progress in improving the country's infrastructure, the government subtracted the decrease from the increase and reported that its actions had increased the capacity of the road by 17 percent. In reality, of course, the capacity returned to what it had been, resulting in no net benefits. The net costs were the price of the paint and an increase in traffic fatalities.

Life is the game that must be played.

E. A. Robinson

13

FUN PROBLEMS

In this chapter, I invite you to leave the real world and enter the world of games and brain teasers. Entering this world can enlighten you, entertain you, and sharpen your thinking. The first problem is one of the oldest of the genre.

The First Night in Paradise

It is the night after Adam and Eve's first day in Paradise. Together, they watched the sun rise and illuminate the marvelous trees, flowers, and birds. At some point the air got cooler, and the sun sank below the horizon. Will it stay dark forever? Adam and Eve wonder, What is the probability that the sun will rise again tomorrow?

With hindsight, we might think that Adam and Eve should have been certain that the sun would rise again. However, they had seen the sun rise only once. What should they expect? The classical answer to this problem is as follows. If Adam and Eve had never seen the sun rising, they would assign equal probabilities to both possible outcomes. Adam and Eve represent this initial belief by placing one white marble (a sun that rises) and one black marble (a sun that does not rise) into a bag. Because they have actually seen the sun rise once, they put another white marble in the bag.

Now that the bag contains two white marbles and one black marble, their degree of belief that the sun will rise tomorrow has increased from 1/2 to 2/3. After observing the sunrise on the following day, they add a third white marble to the bag, reflecting the fact that their degree of belief has increased again, this time from 2/3 to 3/4. Thus, after having seen the sun rise once, Adam and Eve's degree of belief that it will rise again the next morning is 2/3 (Figure 13-1).

According to the *rule of succession,* introduced by the French mathematician Pierre-Simon Laplace in 1812, your degree of belief that the sun will rise again after you have seen the sun rise n times should be $(n+1)/(n+2)$.[1] When you are 27 years old, the sun has risen 10,000 times in your life, so by the rule of succession your confidence that it will rise again tomorrow should be 10,001/10,002.

Unlike in the medical situations in the previous chapters, in which the disease base rates were known, Adam and Eve initially had no base rate for sunrises. In other words, their initial assignment of one marble to each of the possible outcomes was not based on experience. If they had been pessimists, Adam and Eve might have begun with 1 white marble and 10 black marbles in the bag; or, if they had been optimists, they might have done the

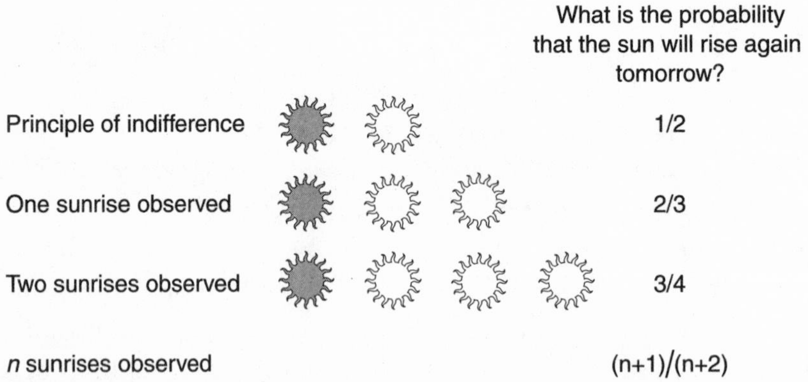

		What is the probability that the sun will rise again tomorrow?
Principle of indifference		1/2
One sunrise observed		2/3
Two sunrises observed		3/4
n sunrises observed		$(n+1)/(n+2)$

FIGURE 13-1. *The principle of indifference.* This principle can be applied when one does not know which of two events (the sun will rise tomorrow or it will not) is more likely to occur. This indifference is represented by one black and one white sun, resulting in a probability of 1/2. When Adam and Eve actually observe a sunrise, one white sun is added and the resulting probability increases to 2/3, and so on.

opposite. The practice of assigning equal probabilities to the possible outcomes when one has no basis on which to estimate their probabilities is known as the *principle of indifference.* It remains controversial. Its proponents defend it by arguing that the initial assignment of probabilities to outcomes has less impact the more observations one makes. For instance, after 10 years of sunrises, one's estimate of the probability that the sun will rise again tomorrow becomes practically the same whether one was a pessimist or an optimist to begin with.[2]

The Base-Rate Fallacy

The story of Adam and Eve enlightens us about making predictions under conditions of almost complete ignorance. The following brain teaser was not meant to enlighten, but to demonstrate the existence of *cognitive illusions* in ordinary humans. It is one of the earliest demonstrations of a cognitive illusion that later came to be called the *base-rate fallacy.* The psychologist M. Hammerton posed the following problem to housewives in Cambridge, England:

1. A device has been invented for screening a population for a disease known as psylicrapitis. 2. The device is a very good one, but not perfect. 3. If someone is a sufferer, there is a 90 percent chance that he will be recorded positively. 4. If he is not a sufferer, there is still a 1 percent chance that he will be recorded positively. 5. Roughly 1 percent of the population has the disease. 6. Mr. Smith has been tested, and the result is positive. The chance that he is in fact a sufferer is: _____ [3]

The housewives failed. Their average estimate of the probability that Mr. Smith suffered from the fictional disease was 85 percent, whereas the correct answer—according to Bayes's rule (introduced in Chapter 4)—is 50 percent. Hammerton suggested that his finding had to do with housewives' lack of experience with medical diagnosis. The real reason, however, may be easier to overcome. The information was presented in conditional probabilities. As I have shown in earlier chapters, there is a simple way to turn

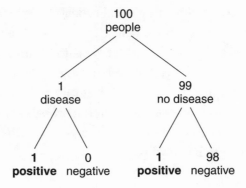

FIGURE 13-2. *Frequency tree for a fictional disease.* A proper representation makes it easy to see that between two people who test positive, only one has the disease.

innumeracy into insight in this problem: Translate the probabilities into natural frequencies (here the frequencies are rounded).

> *Think of 100 people. One has the disease psylicrapitis, and he is likely to test positive. Of those 99 who do not have the disease, 1 will also test positive. How many of those who test positive do have the disease? _____ out of _____.*

The answer is clearly 1 out of 2 (Figure 13-2). The difficulty did not reside in the housewives' inexperience with medical diagnosis. If they had received the information in natural frequencies, the problem would not have clouded their minds in the first place.[4] A proper representation can help to make even the most puzzling "cognitive illusions" disappear.

Why Most Drivers Are Better Than Average

When people are asked how safe their driving is, most respond that they are better than average. Researchers of risk perception have argued that "it is no more possible for most people to be safer [drivers] than average than it is for most to have above average intelligence."[5] The finding that most drivers believe they are better than average has been used as another illustration of people's cognitive illusions, much to the amusement of students of business, psychology, and beyond. It has been variously attributed to people's overconfidence, optimism, or illusion of control—that is, overestimation of their own power to avoid accidents.

Let us have a second look at this phenomenon. It is not possible that most people have above-average IQs because the IQ distribution is, by definition, symmetric. In other words, the number of people with IQs above the average (arithmetic mean) is the same as the number below the mean. Is it possible that most people drive more safely than average? Yes, because safety in driving is not symmetrically distributed around the mean. One can see this result by drawing a frequency distribution. Let me explain.

For illustration, I take the number of car accidents in which a driver is involved over the course of a lifetime as a measure of safety in driving. Imagine 100 drivers. The average number of accidents per driver is 3. If the numbers of accidents were symmetrically distributed around this arithmetic mean of 3, the distribution would look like Figure 13-3 (top). There would be 5 drivers without accidents, 10 with one accident, and so on. The distribution would look the same on both sides of the mean—it would be symmetrical. (The "safe" drivers are shaded in the figure.) Assuming a symmetric distribution such as this one, half of the drivers—and no more—can be better than average.

In reality, however, safety in driving is not symmetrically distributed.[6] A minority of drivers actually have large numbers of accidents. Figure 13-3 (bottom) shows such a distribution, again with 100 drivers. There are a few very unsafe drivers on the right side, and many safe drivers who have had no accidents or one accident on the left side. Rather than being symmetric, this distribution is skewed. The average number of accidents is shifted to the right—to 4.5 accidents—because the few very bad drivers push up the mean. Now it is clear that more than 50 percent of the drivers—63 out of the total of 100—are better than average.[7]

The general point is that a graphical representation can help us to see whether a distribution is symmetric or skewed. When events are distributed symmetrically, 50 percent of cases are above the arithmetic mean and 50 percent below. When a distribution is skewed, as is the case for the num-

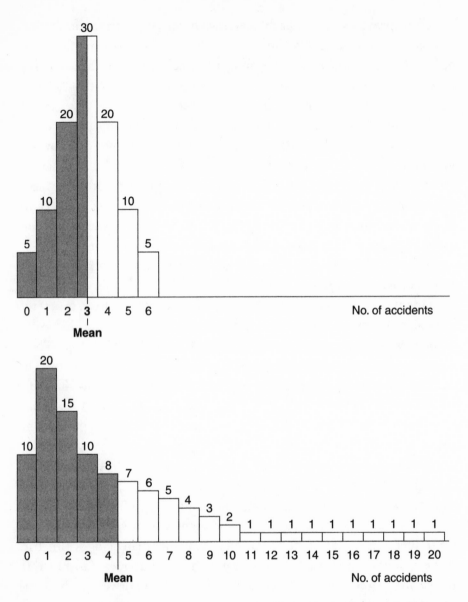

FIGURE 13-3. *Can most drivers be better than average?* When the distribution of accidents is symmetrical (as in the top panel), half of the drivers are better than average (shaded) and half are worse. However, when the distribution is skewed (as in the bottom panel), most drivers *can* be better than average. Here, the mean number of accidents per person is 4.5. Because of the skewed distribution, most drivers—63 out of 100—have fewer than the average number of accidents.

ber of car accidents people have, there is no longer the same number of people on both sides of the mean. The conclusion that people overestimate their driving prowess may be premature. A bit of statistical thinking can show that most drivers actually drive more safely than "average."

The Monty Hall Problem

For about three decades, Monty Hall hosted a popular American game show called *Let's Make a Deal.* His final contestant on each show was given the chance to win a big prize. Here is the situation the contestant faced, as described in *Parade* magazine by the columnist Marilyn vos Savant—allegedly the person with the highest recorded IQ. (Figure 13-4):

> Suppose you're on a game show, and you're given the choice of three
> doors. Behind one door is a car, behind the others, goats. You pick a
> door, say number 1, and the host, who knows what's behind the doors,
> opens another door, say number 3, which has a goat. He says to you,
> "Do you want to pick door number 2?" Is it to your advantage to
> switch your choice of doors?[8]

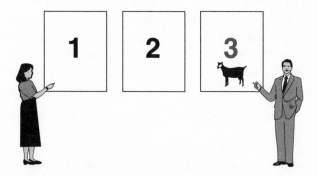

FIGURE 13-4. *The three-doors problem.* The guest on the Monty Hall show has a choice among one of three doors. Behind one is a car, and behind the others are goats. The guest (left) picked door 1. Monty, who knows where the car is, opens door 3 and shows a goat. Should the guest switch to door 2 or stay with 1?

What would you do? Stay or switch? I asked a graphic designer this question. Our conversation appears below.

Designer: I would stay.

GG: Why?

Designer: You should not revise a decision you've made.

GG: Why?

Designer: I would feel bad if the car were behind the door I chose and then I had switched to one with a goat.

GG: You would feel worse if you switched from a door with a car than if you stayed with a door with a goat, although you would lose in both cases?

Designer: Definitely.

GG: Now imagine there were 100 doors with 99 goats and one car. You choose door number 1, and Monty opens all the doors except number 37, and each of the doors he opens shows a goat. Would you switch to number 37?

Designer: No. I would stay. There are two doors closed, and the chances are 50-50. No reason to switch.

GG: The chance of choosing the right door is 1 in 100, isn't it? So, in 99 out of 100 cases, you will have picked a door with a goat, and Monty will show you all the other doors with goats, leaving the door with the car closed. In these 99 cases, staying loses and switching wins. In only one case, the case in which you are lucky enough to have picked the door with the car, is switching to your disadvantage.

Designer: I'm getting confused. I still would not switch.

The vast majority of people think, as the graphic designer did, that switching and staying are equally likely to win them the car—and decide to stay.[9] But vos Savant recommended that the contestant switch doors. Her solution to this brain teaser caused a nationwide debate. *The New York Times* featured a front-page article that explored the fuss about switching or not switching, and thousands of opinion letters swamped *Parade* and other magazines.[10] The authors of most of the letters insisted that Marilyn was wrong. A professor of mathematics wrote the following letter:

Dear Marilyn:

Since you seem to enjoy coming straight to the point, I'll do the same. In the following question and answer, you blew it! Let me explain. If one door is shown to be a loser, that information changes the probability of either remaining choice, *neither of which has any reason to be more likely,* to 1/2. As a professional mathematician, I'm very concerned with the general public's lack of mathematical skills. Please help by confessing your error and in the future being more careful.

Another writer got down to the very essence of the problem:

You cannot correctly apply feminine logic to odds. The new situation is one of two equal chances.

In a final illustration, a college professor defends three of his colleagues who were all of one opinion with him:

. . . As an educator, I find your . . . column reprehensible, and urge you to reconsider your future endeavors. In my opinion, you no longer provide a useful service to your readers. . . . I am encouraging *Parade* magazine to drop your column. I would also encourage your publisher to print a strong apology to the three scholars whose wisdom was impugned by your illogical babble. I hope they sue you!

As far as I know, the scholars never sued vos Savant, and they were well advised not to. To switch or not to switch—that is the question. To answer it, we have to make a few assumptions that were left out of vos Savant's version of the Monty Hall problem. The first assumption is that Monty always gives guests a chance to switch, or at least that whether he gives them a chance to switch does not depend on which door they choose. For instance, if he offered a chance to switch only when the guest picks the door with the car, then not switching is obviously the winning strategy. The second assumption is that Monty always opens a door with a goat, never the door with the car. The third assumption is that Monty makes all other relevant choices randomly, including the choice of behind which door to place the car and the choice of which door to open when he has a choice between two doors with goats. If the situation is clarified in this way, then one can show that vos Savant was right: Switching increases the contestant's chance of winning from 1/3 to 2/3. However, as in the previous problems, one can see the answer best with the help of a proper information representation. There are several representations that illustrate that switching is to the contestant's advantage.[11]

> *Frequency judgments.* Don't ask whether you should switch or stay, but in *how many cases* switching pays. In other words, don't ask about a single event, but rather ask about repeated events. This frequency question reveals that—at the point where you are deciding whether to switch—there are three possible cases: (1) You have picked a door with a goat, (2) you have picked the door with the car, or (3) you have picked the other door with a goat. If you don't switch, in only one out of the three cases—the second one—will you win the car (Figure 13-5). If you switch, you will win the car in two out of the three cases. This occurs in the two cases in which you picked a door with a goat, Monty opened the other door with a goat, and you switched to the remaining door—the one with the car.
>
> *Perspective change.* Picture yourself in the role of Monty rather than in the role of the guest. In other words, imagine standing behind the doors rather than in front of them. As the host, you know where the car is. Assume the car is behind door 3 and the guest has already made her first choice. Three scenarios are possible. If the guest picked door 1, you are forced to open door 2. If she now switches, she will win the car. If the guest picked door 2, you are forced to open door 1. If the guest now switches, she will win the car. If the guest picked door 3, you will randomly open either door 1 or door 2. Out of the three scenarios, only in this last one will

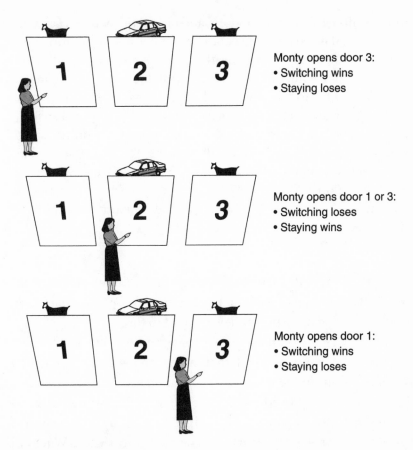

Monty opens door 3:
• Switching wins
• Staying loses

Monty opens door 1 or 3:
• Switching loses
• Staying wins

Monty opens door 1:
• Switching wins
• Staying loses

FIGURE 13-5. *Does switching pay?* The guest can choose either door 1, 2, or 3. If she stays with her choice of door 1, she will win the car in only one out of three cases. If she switches, she will win the car in two out of three cases.

switching result in the guest's not winning the car. In short, a guest who switches will win in two out of the three cases, and a guest who stays will win in only one of the three cases. Through Monty's eyes, it is easier to see that switching increases the chance of winning from 1 in 3 to 2 in 3.

Repeated play. A third way to see that switching pays is to simulate playing the game repeatedly. Take three coffee cups, turn them upside down, choose one at random, and put a $10 bill underneath it. Ask a contestant to pick one cup. Then let her watch you turn over a cup with no money underneath it, and proceed as Monty Hall would. Your friend can either stay with the cup she originally picked or switch to the other cup that has not yet been turned over. Repeat this game 100 times with a contestant who does not switch, and observe how many $10 bills she collects. Then do the same with a contestant who switches all the time. You will observe that the person who never switches will win about $300 to $350, that is, about one-third of the games. In contrast, the person who always switches will win about $600 to $700, that is, about two-thirds of the games.

The bottom line is this: The best strategy in the Monty Hall dilemma is to switch. Those who don't switch leave money on the table. Lessons learned by winning or losing money are hard to forget.

So it seems that we should advise a contestant on *Let's Make a Deal* to switch. But wait a moment. Recall that this chapter is about fictional problems, not problems in the real world or even on game shows. Switching is the best strategy in the word problem above, with the assumptions added, but the real game show may have had different rules. Take the assumption that Monty always gave the final guest a chance to switch. When asked whether this was true, the real Monty Hall recalled that he rarely offered guests the switching option and could not say how often it was accepted. However, his longtime production assistant, Carol Andrews, asserted that Monty never offered the last guest an opportunity to switch. Barry Nalebuff, who wrote one of the first articles on the Monty Hall problem, says that in watching the show he saw the switch option offered, but he cannot recall whether Monty made the offer in every show or whether his making the offer depended on which door the guest had picked in the first place.[12] In the real world, even game shows are uncertain. And in this case, the uncertainty is not merely a product of imperfect memory. The real game show revolved around Monty Hall, and part of his personality was to make spontaneous decisions rather than to follow a strict protocol. In other words, part of the suspense of the game arose from contestants' uncertainty about Monty Hall's motives and actions, and this suspense might have been lost if he had followed the same rules in show after show.[13]

Three Prisoners on Death Row

In a country far away, three prisoners—Tom, Harry, and Dick—are awaiting execution in separate cells. They have just heard the news that a healthy daughter has been born to the king and that the king has decided to express his gratitude to God by pardoning one of the prisoners. Because the monarch does not care which prisoner is spared, the lucky one has been chosen by lottery. Each prisoner has a 1/3 chance of being pardoned. The warden, who already knows the result of the lottery, has been instructed not to let any of the prisoners know if he is the one who will be pardoned. Arguing that everyone knows for sure that either Tom or Harry will be executed (because only one or the other—or neither—of them will be spared), Dick persuades the warden that it would not be a violation of his instructions to tell Dick whether Tom or Harry will be executed. The warden tells Dick that Harry will be executed. Dick cheers up. Either he or Tom is the one who will be pardoned, so he concludes that his chance of survival has increased from 1/3 to 1/2. Is his reasoning correct?

The story of the three prisoners usually generates lively discussion.[14] Listen to two students consider the question:

Ilona: Two prisoners are left; therefore each has a 50-50 chance. It's that simple.

Lara: No, Dick still has a 1/3 chance. How can Dick's chance increase?

Ilona: Count to two! At first, there were three guys, and the chance was 1/3 for everyone; now two guys are left, which makes 1/2 for each. If there were only one guy left, his chance would be 1. It's that simple.

Lara: No, I don't think so. Dick has still a 1/3 chance—nothing more. The warden did not tell him anything about himself; he knows nothing new about his situation. No news, no change.

Ilona: Come on, that doesn't make sense. Just think about what you're saying. If nothing changed, then Dick would keep his 1/3 chance, and Tom his 1/3 chance, and since Harry lost his 1/3, there is something missing. Just use your brain. Chances can't disappear!

Lara: They don't—Tom has now a 2/3 chance of being pardoned.

Ilona: You've got to be kidding me. That's not fair.

Lara: Stop driving me crazy. Dick has a 1/3 and Tom a 2/3 chance. You think that because the warden named Harry, Dick's chances increased to 50-50. Look, if the warden had named Tom, you also would think that Dick's chances increased to 50-50. Whatever the warden says, you think that Dick's chances increase. Dick doesn't even have to listen to the warden, because his chances increase anyway! The warden could be deaf! Or mute. Or Dick could just dream of the warden and his chances would still miraculously increase to 50-50 because the warden's answer doesn't matter! You can't believe such nonsense, can you?

Ilona: Calm down, you are getting emotional. Just use your brain. Imagine what would have happened if Tom had asked the warden. Whatever the warden answers, with your argument, Tom's chance would remain 1/3. But you can't tell me that Tom's chance depends on who asked the warden. You argue that if Dick asked, Tom's chance would be 2/3; but you tell me that if Tom asked, Tom's chance would be only 1/3. That's not logical.

Lara: No, it is—the problem is in your story. You tell me that merely asking the warden makes a prisoner's chance of pardon increase to 1/2— whatever the warden's answer is. You can't make anyone believe this. If Tom, Dick, and Harry—I mean, each and every one of them—asked the warden, then you're telling me that each one would return to his cell thinking that his chance is 1/2. All three are happy! You're telling me that's logical?

Do Dick's chances increase to 50-50 or not? To answer this question, we need to spell out some assumptions, just as we did in the Monty Hall problem. First, the warden would not tell Dick that he is the one to be spared (just as Monty Hall would not open the door with the car); and second, if the warden has a choice between naming Harry or Tom—that is, in the event that Dick is the one to be pardoned—the warden would make the choice randomly (just as Monty Hall would randomly choose which of the two doors with goats to open when the guest had already picked the door with the car). Given these assumptions, Lara is right: Dick's chance has not changed in light of what the warden told him—it is still 1/3. The reason is the same as in the Monty Hall problem: The probability that the car is behind the door that the contestant has chosen does not change after Monty Hall opens a door with a goat. Similarly, the probability that Dick will be pardoned does not change after the warden reveals the name of a prisoner who will be executed.

The guest on *Let's Make a Deal* corresponds to Dick, Monty Hall to the warden, winning the car to being pardoned, and the opening of a door with a goat to the naming of a prisoner who will be executed. If Dick could metamorphose into one of the other men, he should chose to become Tom because Tom has a better chance of being the lucky one. To better understand this, let's take a look at the same representations we used in the Monty Hall problem.

Frequency judgments. In the Monty Hall problem, we recast the question of whether the guest should stay or switch as how often switching pays. Similarly, in the three prisoners problem, we can recast the question of whether asking the warden "pays" to how often asking pays. As in the Monty Hall problem, this question focuses attention on the three possible cases: (1) Tom has been pardoned, (2) Harry has been pardoned, and (3) Dick has been pardoned (Figure 13-6). Initially, each prisoner's probability of being spared is 1/3. If Tom is to be pardoned, the warden tells Dick that Harry will be executed. In this case, Dick will also be executed. If Harry is to be pardoned, the warden tells Dick that Tom will be executed. In this case, Dick will also be executed. If Dick is to be pardoned, the warden tells Dick (at random) either that Harry or that Tom will be executed. Only in this case will Dick be spared.

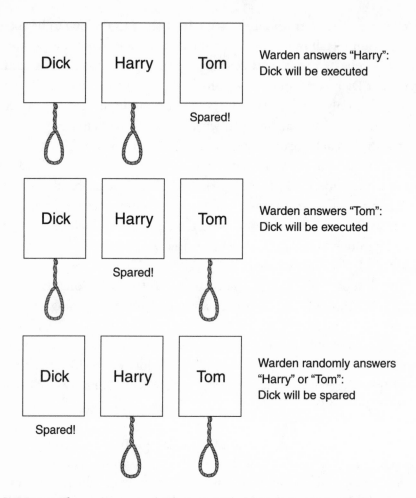

FIGURE 13-6: *Three prisoners on death row.* Is Dick right in believing that his chances of being spared have increased from one-third to one-half? There are three possibilities: Either Tom, Dick, or Harry will be spared. In two of the three cases, Dick will be executed, whatever the warden tells him. Thus, Dick's chances of being spared do not increase, but the chances of the person whom the warden did not name increase from one-third to two-thirds.

Now we can see that Dick will be executed in two out of the three cases. In other words, his probability of being pardoned remains 1/3, even after the warden answers his question.

Perspective change. Picture yourself in the role of the warden rather than in the role of Dick. You know who is going to be freed. Now one of the prisoners asks you to name one of the prisoners who will be executed. You know that in two out of three cases, the prisoner who asks will be one of the prisoners marked for execution. For instance, assume that all three prisoners ask you to name one of the prisoners who will be executed. Two of them will be marked for execution. Through the warden's eyes, it is easier to see that after asking the warden, the prisoner's chance of being pardoned is still 1/3.

Repeated play. A third way to understand the problem is to play the game repeatedly. Take three coffee cups, each marked with the name of one of the three prisoners, and randomly choose one cup under which to put a slip of paper that says "Spared." Playing the role of the warden, now turn over one of the cups that does not have the slip of paper under it—that is, reveal that either Harry or Tom will be executed, just as described above. Repeat this game 100 times and see how often "Spared" is underneath the cup marked "Dick." You will observe that Dick is spared in about 1/3 of the cases, not in 1/2 of the cases.

Playing with Representations

Many of us have been tricked by fun problems. In retrospect, we wonder why we did not see the solution before. Just as with the real problems discussed in the previous chapters, proper representations can facilitate insight. Fun problems can be a motivating start to learning how to reckon with risk by playing with representations. Because of their relative simplicity, fun problems can also easily clarify the assumptions one needs to make to actually determine an answer. In the next and final chapter, I will deal with the larger issue of teaching thinking, and I now invite you to reenter our modern, technological world.

I know no safe depository of the ultimate powers of the society but the people themselves; and if we think them not enlightened enough to exercise their control with a wholesome discretion, the remedy is not to take it from them, but to inform their discretion.

Thomas Jefferson, Letter to William Charles Jarvis, 1820

14

TEACHING CLEAR THINKING

In the United States, an estimated 44,000 to 98,000 people die in hospitals each year because of documented and preventable medical errors.[1] In Germany, between 8,000 and 16,000 patients in hospitals die every year because they are administered the wrong medication or the right medication in the wrong amount, and several hundred thousand others develop serious illnesses as a result of such errors.[2] An unknown number of people who are not infected with HIV commit suicide because they test positive in screening, not realizing that false positives occur. Each year, some 100,000 German women without breast cancer have part of their breasts surgically removed in the follow-up after a positive test, and only a few of them know that most positive screening mammograms are false positives. As mentioned in a previous chapter, after learning of his own prostate cancer, Mayor Rudolph Giuliani of New York is reported to have urged "everyone to get the PSA test."[3] Thousands of men have followed this advice, many of them not knowing there is no evidence that early detection of prostate cancer reduces mortality. These men can expect no benefit from treatment—only the risk of its side effects, which include incontinence and impotence.[4]

In the courts, jurors and attorneys struggle with random match probabilities, which lead to confusion and even convictions that later had to be revised.[5] The list goes on. The mind tools presented in this book can contribute to reducing unnecessary anxiety and stress and to saving lives.

How can we prepare the next generation to understand the risks it will face? Our children will live in a world with new technologies, including tests for many genetic diseases. The information afforded by these technologies, and the risks that their use implies, need to be understood—in terms of both benefits and costs. At present, high school education in some countries covers only little, if any, statistical thinking. Algebra, geometry, and calculus teach thinking in a world of certainty—not in the real world, which is uncertain. Medical schools routinely teach statistics, but focus on methods such as significance tests rather than the kind of statistical thinking needed for sound diagnosis and risk assessment, as the widespread innumeracy in the medical profession, documented in Chapters 5, 6, and 7, attests. Furthermore, in the medical and social sciences, data analysis is typically taught as a set of statistical rituals rather than a set of methods for statistical thinking.[6] The situation in the law profession seems to be even worse. With one or two notable exceptions, law schools do not teach students how to reason on the basis of uncertain evidence—although virtually all evidence is uncertain. The mind tools described in this book are mostly unknown in all these disciplines.

The time is ripe for an educational campaign aimed at teaching schoolchildren, undergraduate and graduate students, ordinary citizens, and professionals how to reckon with risk. What would a curriculum for teaching thinking in an uncertain world look like? The three steps elaborated below outline an educational program that can be adapted to any specific discipline or level of study. The first step is to teach people to recognize types of uncertainties, including uncertainties disguised as certainties. The second step is to teach turning uncertainty into risk—that is, estimating degrees of uncertainty. The third step is to teach people to experiment with representations so as to discover the most transparent way to communicate and reason with risks. All these steps have been described in this book; in this final chapter, they are integrated into a single program for fostering statistical insight.

Step One: Franklin's Law

Recall Benjamin Franklin's aphorism "In this world nothing is certain but death and taxes." This statement reminds us that almost all real-world events are uncertain and that we need to learn to deal with—rather than deny—this fact. Giving up the illusion of certainty enables us to enjoy and explore the complexity of the world in which we live. The first step in the educational program is to make people aware of Franklin's law. The flip side of this goal is to reduce the illusion of certainty. The tools for taking the first step include (1) real stories that illustrate the uncertainty inherent in everyday situations and (2) causal understanding of uncertainties and errors. The best educational material is found in real events, not in hypothetical examples. Described below is a true story of an extreme instance of certainty turning into uncertainty.

35 NEGATIVE TESTS

In April 1995, a previously healthy 36-year-old American construction worker suffering from fatigue was tested for HIV.[7] The ELISA, a test for HIV approved by the U. S. Food and Drug Administration, returned a negative result. Two months later, the man, who had lost 27 pounds, was admitted to a hospital with shortness of breath, diarrhea, and other symptoms. A second ELISA test indicated that he was HIV-negative, and routine laboratory tests turned up no other illness. The patient was discharged without a diagnosis. In August of the same year, the man was hospitalized again. An ELISA and a Western blot performed by the Utah Department of Health laboratory came back negative. At that point, the physicians decided to interview his wife, from whom he had separated two years previously. She reported having had sexual contact with an HIV-infected partner before their marriage and told the doctors that this earlier partner had recently died of AIDS. In 1994, she developed pneumonia and tested positive for HIV, a fact of which the construction worker was unaware. He reported that during their marriage, he had had sex with his wife without using a condom, but he had had no sexual contact with her since their separation. Because of his history of exposure and the strong clinical evidence that his immune system was compromised, the doctors performed a series of additional laboratory tests, which eventually revealed that he had the same strain of HIV as his ex-wife.

During his examination, the construction worker made it known that he had—in good faith—donated blood more than 30 times in the previous four years. In each case, the routine ELISA—which is used to screen all blood donations for the HIV virus—returned a negative result. The consequences for those who received the construction worker's blood are unknown. This HIV-infected man had tested negative for HIV no fewer than 35 times over a four-year period.

How can this incredibly long series of false negative results—which may have had devastating consequences for recipients of the man's donated blood—have occurred? To begin with, we do not know whether all the man's negative results were false negatives, because the point at which he became infected is not known. He continued to receive false negatives, however, even after he developed clear signs of severe immunosuppression, including the opportunistic infections and low lymphocyte counts characteristic of AIDS. There are two main reasons why HIV tests sometimes return false negatives. First, there is a period—usually about six months—after the time of infection during which HIV antibodies cannot be detected. But the construction worker's test results were negative far beyond the usual window period. Second, new strains of HIV, which arise from the virus's ability to mutate quickly (see Chapter 7), cannot always be detected by routine HIV tests. But the strain of HIV carried by this patient was practically identical to that carried by his wife, who had tested positive.

Why the construction worker tested negative for HIV despite having been infected has yet to be explained. A case like this is very rare; nevertheless, it could happen again. The ELISA and other HIV tests are among the best antibody tests ever devised, yet certainty remains out of their reach.

THE ILLUSION OF CERTAINTY

This extreme case illustrates how broadly Franklin's law applies. Starting in Chapter 1, we encounter numerous cases of illusory certainty in this book. Many people, of varying levels of education, believe that the results of HIV tests, DNA fingerprinting, and the growing number of genetic tests are absolutely certain. These technologies are formidable, but not foolproof. Like just about everything else, they are subject to Franklin's law.[8]

Step Two: Beyond Ignorance of Risks

Why do American boys know the batting averages, earned-run averages, and won-lost records of their home baseball teams, while most of their adult counterparts are largely ignorant of the statistics that describe the world outside the baseball stadium, such as the number of Americans killed by handguns every year? How can European boys know the record of their favorite soccer team for the last several years, while most of their adult counterparts have no idea what their chance is of being killed driving on the highway? Why are most women unaware of the benefits and costs of breast cancer screening, and why are most men unaware of those of prostate cancer screening?

John Q. Public's ignorance of risks is not entirely his fault; it originates not only inside but also outside his mind. Internal sources of ignorance include a preference for distraction over information or for passivity over responsibility. But ignorance of risks is also fueled by external factors ranging from peer pressure to lobbying by trade associations. Step Two in the program of teaching thinking entails overcoming both internal and external sources of ignorance. The goal is (1) to teach people how to use tools for estimating risks, including the uncertainty around these estimates, and (2) to make people aware of the forces aimed at preventing them from estimating risks.

"DOUBT IS OUR PRODUCT"

The United States is home to thousands of trade associations promoting everything from asbestos to zinc. The Beer Institute defends brewers against claims that drunk driving causes car accidents. The Asbestos Information Association protects citizens from their "fiber phobia." The Global Climate Coalition represents scientists who question the evidence for global warming. Washington, D.C., alone, is home to 1,700 such trade as-sociations. Estimates indicate that more than $1 billion is spent by such or-ganizations every year on "image advertising" and "issues management." Trade associations have become active in the manufacture of knowledge and ignorance. Consider, as an example, the Tobacco Institute's "spin" on the hazards of cigarette smoking.[9]

At the beginning of the twentieth century, lung cancer was an excep-tionally rare type of cancer—so rare that Isaac Adler, who wrote the first book-length medical review on lung cancer, apologized for writing about such an uncommon and insignificant disease.[10] By the end of the twentieth century, lung cancer had become the most frequent cause of cancer deaths worldwide. Why? At the beginning of the century, cigarette smoking was rare; people smoked pipes and cigars. Smoking cigars causes different kinds of cancer than smoking cigarettes does. To take an example, Sig-mund Freud developed cancer of the mouth as a result of his heavy cigar smoking. The cancer cast a shadow over the last 16 years of his life, causing him continuous pain and discomfort and requiring some 30 operations to remove cancerous and precancerous growths.

Cigarettes first became popular during World War I. Unlike cigar and pipe smoke, cigarette smoke is generally inhaled, exposing lung tissue to ir-ritants. The link between cigarette smoking and lung cancer was first demonstrated by German researchers in the 1920s and 1930s but was largely ignored in America, possibly because this research was associated with the Nazis. In the early 1950s, however, a consensus developed in the American scientific community that cigarettes are a major source of illness, including lung cancer. By the mid-1950s, there was strong evidence that a two-pack-a-day smoker lived, on average, about seven years fewer than a nonsmoker. Most scientists came to agree that tobacco kills about 400,000 Americans every year and that tobacco is the cause of 80 percent to 90 of lung cancers.[11]

The Tobacco Institute was founded in 1958 as an offshoot of the Coun-cil for Tobacco Research, which was established by tobacco manufacturers, growers, and warehouse owners. Since then, it has argued the case for cig-arette "safety" by creating doubt in the public mind about the hazards of smoking. In the 1960s, spokesmen tried to undermine and distract from the growing consensus in the scientific community. For instance, they asserted that the link between cigarettes and cancer was "merely statistical," that the evidence was uncertain and the conclusions premature, and that there might be a gene that both leads to smoking and predisposes certain people to developing cancer. In 1962, a Gallup survey found that only 38 percent of American adults knew that cigarettes cause lung cancer. Although many physicians quit smoking after the Surgeon General's Report in 1964 made it clear that cigarettes are a major cause of illness, many members of the public remained under the impression that the question about the effects of smoking on health was still open. The silence in popular magazines about the hazards of smoking played a crucial role in maintaining public ignorance; cigarette advertisers discouraged magazine publishers from covering the topic of smoking hazards. A 1978 article in the *Columbia Journalism Review* noted that it could not find a single article in a leading national magazine that had discussed the health effects of smoking in the last seven years. The less sophisticated, popular press was more straightforward. A headline in a 1968 *National Enquirer* read: "Most Medical Authorities Agree, Cigarette Lung Cancer Is Bunk: 70 Million Americans Falsely Alarmed."[12] Much later, in 1989, the Surgeon General's Report explicitly linked the tobacco lobby's suppression of media coverage to the general public's ignorance of the nature and extent of the hazards of smoking.

More recently, the Tobacco Institute has tried to challenge evidence of the hazards of "passive smoking" or "secondhand smoke." Strong evidence of the negative health effects of breathing smoke from others' cigarettes emerged in the 1980s, when Tokyo's National Cancer Center Research Institute showed that lung cancer was twice as common among the nonsmoking wives of smokers as among those of nonsmokers.[13] In the 1990s, the Environmental Protection Agency released data indicating that secondhand smoke was responsible for 20 percent of all lung cancer deaths among nonsmokers, that is, for the deaths of about 3,000 Americans a year. The Tobacco Institute dismissed this study as "characterized by a preference for political correctness over sound science."[14]

The case of the tobacco lobby epitomizes the manufacture of ignorance and confusion. Its efforts at obfuscation shifted constantly; as soon as one argument from the tobacco lobby was discredited, new arguments were constructed to engender fresh confusion. Their arguments and slogans evolved in the following way:

Smoking doesn't hurt your health; it's safe.

OK, smoking may or may not hurt your health, but the scientific evidence is still insufficient and inconclusive.

OK, the evidence is conclusive that smoking does cause lung cancer, but we didn't know until now.

OK, we knew, but we didn't know that nicotine was addictive.

OK, we knew that nicotine was addictive when we added chemicals to cigarettes to make nicotine enter the bloodstream faster, but this was long ago. Today we have low-tar and low-nicotine cigarettes.

OK, low-tar and low-nicotine cigarettes do not actually reduce the risk of lung cancer, but this is people's own fault because they now smoke more cigarettes.

OK, it is in our interest that people smoke more, but they smoke more by their own free choice.

A similar sequence of claims has been made to cloud people's minds concerning the risks of passive smoking. As the historian Robert Proctor reports, this goal was privately admitted in an internal document produced by a cigarette company: "Doubt is our product since it is the best means of competing with the 'body of facts' that exists in the mind of the general public."[15]

WHAT DOES JOHN Q. PUBLIC FEAR?

A few years ago, I booked a flight across the Alps from Munich to Florence. Entering the small Italian airplane with a boarding card labeled seat 14A, I walked down the narrow aisle, squinting to read the row numbers. When I had almost reached the end of the aisle, I saw row 12, after which there was only one more row. I thought I was in the wrong plane. Then I noticed that the last row was row 14 and that there was no row 13. A light went on in my head: The airline had skipped row 13 in deference to Europeans' super-

stition that the number 13 is a bad omen, much as the designers of eleva-
tors in many American skyscrapers know to "skip" the 13th floor. In India,
where this superstition is unknown, 13 is treated just like 12 and 14.[16]

What John Q. Public most fears is not always what threatens him most.
Psychological research has identified, among others, the following three
sources of fear.[17]

Preparedness. Fear of natural, recurrent threats that have endangered us throughout our evolution is easy to learn, whereas fear of
evolutionarily novel threats is often hard to learn. This ability to
learn to fear an object from only one or very few observations is
called *preparedness.* Rhesus monkeys reared in the laboratory, for
instance, show no fear of venomous snakes. However, when a
youngster watches an adult exhibiting fear of a snake, the youngster typically acquires this fear just by observing it once. The logic
behind this genetic preparedness to learn some things faster than
others is obvious. If the youngster had to learn from experience
that a snake is poisonous, its chance of survival would be slim.
Hence, evolutionary learning accelerates individual learning. But
preparedness holds only for certain stimuli. For instance, when the
youngster sees another monkey exhibiting a fear reaction towards
a flower, it does not acquire a fear of flowers.[18] Humans show similar preparedness for learning. It is easy to get a child to fear spiders, snakes, and tigers. All that is needed is that a parent express
fear in the presence of a spider and that the child observe the parent's fear. It is hard to get a child to fear electrical outlets. Yet, in
modern, industrialized societies, children are much more likely to
be harmed by electrical outlets than by spiders. Fears and phobias
tend to settle on stimuli that have been dangerous in the past.
Darkness, to provide another example, is something that we do not
need to teach our children to fear. But humans have changed their
world dramatically in the evolutionarily recent past. This is one
reason why the things that we fear most are not necessarily the
ones most likely to hurt us.

Disasters. People tend to fear situations that can take many lives at
once. Situations that cause the same number of deaths spread over

time tend to be less feared.[19] For instance, the causes of death that people tend to fear most, such as plane crashes and nuclear accidents, often have catastrophic potential. Car accidents and cigarette smoking, in contrast, cause a continuous stream of deaths, and despite their having killed many, many times more people than plane crashes or nuclear power accidents, they do not evoke similar levels of fear. Like preparedness for learning, the fear of disasters has an evolutionary rationale. When a group drops below a certain size, the group may be extinguished. When a loss of the same size is distributed over time, however, the community or species may be in a better position to compensate and survive.

Unknown hazards. People tend to fear dangers that are new and unknown. Examples include genetic engineering and nuclear technology, as opposed to drinking alcohol. When the new and potentially dangerous technology is additionally perceived to be in the control of unfamiliar people such as the rulers of a foreign country, fear tends to skyrocket.

In sum, John Q. Public does not always fear the situations that are actually most likely to hurt or kill him and other people. The objects and situations we tend to fear are often those that were dangerous in our evolutionary past—such as snakes, spiders, large cats, darkness, being alone, and being exposed in an open place—though most of them are no longer the greatest threats in our modern technological world.

HELPFUL SOURCES

Many sources can provide information to help us find out the risks associated with specific behaviors and courses of action. For instance, the National Safety Council publishes an annual booklet, *Accident Facts,* that lists all the ways in which Americans die accidentally according to the frequency with which they occur. The National Research Council has a series of publications on matters such as understanding and preventing violence[20] and the evaluation of DNA evidence.[21] The *Guide to Clinical Preventive Services* by the U.S. Preventive Services Task Force informs the public on questions about health and medical screening,[22] and so does the *Canadian Guide to Clinical Preventive Health Care* by the Canadian Task Force (see www.ctfphc.org). Noncommercial groups of physicians, such as the Cochrane Centers (see www.cochrane.org) and Bandolier at the University of Oxford (see www.jr2.ox.ac.uk/bandolier), distribute the information patients need to know over the Internet. The *Dartmouth Atlas of Health Care* offers information about the puzzling variations in rates of common surgical procedures performed in different hospital referral regions in the United States.[23] A number of other relevant sources are available in any library. A good source not only gives the risks, but also indicates the uncertainty involved in estimating the risk. Many potential hazards, however, such as the possible link between brain cancer and cellular phones, are new, and scientific studies are few. With respect to these hazards, we continue to live in the twilight of uncertainty.

Step Three: Communication and Reasoning

Information needs representation. The idea that it is possible to communicate information in a "pure" form is fiction. Successful risk communication requires intuitively clear representations. Playing with representations can help us not only to understand numbers (describe phenomena) but also to draw conclusions from numbers (make inferences). There is no single best representation, because what is needed always depends on the minds that are doing the communicating. If you want to inform someone about false positives in screening, what constitutes a "good" representation depends on whether the person is a statistician, a physician, or the patient being examined.

With frustration, one of the physicians who tried in vain to determine the probability that a person with a positive screening test actually has colorectal cancer (Chapter 6) exclaimed: "I know there is a rule; I learned it in school, but I have forgotten it." He was referring to Bayes's rule. In this book, we have encountered many situations in which Bayes's rule can help one to draw conclusions from numbers, such as how to estimate the chance of a disease given a positive diagnostic test result. And we have seen that this type of inference is difficult for laypeople and experts to make when numbers are represented in probabilities, yet comparatively easy when they are represented in natural frequencies. Representation matters to statistical thinking.

One of the goals of formal education that has been neglected in most Western countries is to teach people to reckon with risk, that is, to reason effectively in an uncertain world. One tool for reaching this goal—without engendering "math anxiety"—is to use intuitively understandable representations of numbers. The didactic power of representations does not seem to be widely recognized in the classroom. Consider, for instance, the usual method of teaching Bayes's rule in German high schools. Bayes's rule is taught in some, but not in all, German states. (Many of the teachers my colleagues and I have interviewed believe that Bayes's rule is not particularly important because it is not part of the nationwide high school curriculum.) German-language textbooks that include Bayes's rule explain it exclusively in terms of probabilities or relative frequencies; natural frequencies, which would help students understand the problems better, are never used.[24] Similarly, a survey of teachers revealed that in the classroom almost all of them express risks in terms of probabilities and percentages, and very few teachers tried to foster insight by using natural frequencies. This situation is worsened by the fact that the examples used to introduce Bayes's rule are usually, from a teenager's point of view, thoroughly boring (such as the standard problem of having to guess the probability of there being a golden coin in a particular drawer).

I have heard adults argue that teenagers are not much interested in statistics and that their lack of motivation accounts for our failure to educate them in statistical thinking. However, there is clear evidence to the contrary. In one study, German mathematics teachers reported that, when it comes to statistics, there is a striking discrepancy between students' motivation and performance. They observed that students' interest in statistics is considerably higher than in other mathematical topics, as is their attention and motivation. Sadly, however, the teachers also reported that students' performance was at a considerably lower level in statistics than in other areas of mathematics. This discrepancy indicates that what is lacking is not motivation on the part of students, but rather adequate tools for helping students gain insight into uncertainty and risk—such as intuitively understandable representations of numbers.

COMPUTER TUTORIALS

One class of tools for teaching thinking is computerized tutorial programs. Peter Sedlmeier and I designed such a program to teach people to draw conclusions from numbers. The program instructs the student to solve problems—such as the medical and legal problems described in this book—by translating probabilities into natural frequencies. We call this "representation training."[25] More specifically, the tutorial teaches the student to construct frequency representations by means of trees (as in Figure 4-2, page 45). Its goals are to improve people's performance in the short run and to keep them from forgetting how to solve such problems in the long run.

We compared this representation training with traditional "rule training," in which the learner is taught how to insert probabilities into Bayes's rule. Both types of training were implemented as computer tutorials. Each has two parts. The first part guides the learner through two training tasks step by step: inferring the presence of cancer from a positive test and inferring the presence of sepsis from a high fever, chills, and skin lesions. The representation training program showed participants how to transform probability information into a frequency tree. The rule-training program instructed learners how to insert probability information into Bayes's rule. In the second part of the tutorial, the learner is asked to solve eight additional problems with the help of step-by-step feedback. If participants have difficulties or make mistakes, both programs provide immediate assistance or feedback. The help was sufficient to ensure that all participants could complete all the steps of each problem correctly and complete the training.

Which was more effective: teaching people how to insert probabilities into Bayes's rule (rule training) or teaching them how to translate probabilities into natural frequencies (representation training)? What is effective in the short run need not be effective in the long run. Witness the frustration of devoted teachers who observe that, after an exam, students forget the material at a faster rate than they learned it. However, if natural frequencies touch a chord in the human brain that has been there all along, students who learn a concept in natural frequencies should forget it more slowly than students who learn the same concept in probabilities.

We evaluated the tutorials with respect to both the immediate learning effect and its stability over time. Note that all test problems were expressed in probabilities in both types of training. Both tutorials also allowed participants to work at their own pace. Figure 14-1 shows the results of two studies, one with students from the University of Chicago (labeled "American students"), the other with students from the University of Munich (labeled "German students"). The American students took between one and two hours to complete the tutorials, including the tests before and after training; the German students took slightly more time.[26]

The performance of participants in both studies was very low before

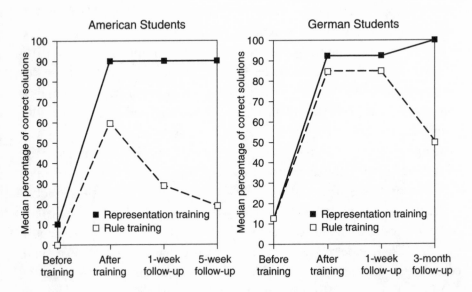

FIGURE 14-1. *How quickly do students forget what they have learned?* With the traditional method of teaching people how to insert probabilities into Bayes's rule (rule training), both groups of students tend to forget what they have learned. However, when they have been taught to use the mind tool of representing probabilities as natural frequencies (representation training), performance remains at a high level.

training. After the representation training, the percentage of correct answers in the American study increased from 10 to 90 percent, whereas after the rule training, it increased from 0 to 60 percent. For the German students, who performed at a slightly higher level before training, the effect of the representation training was of comparable size. Overall, both tutorials showed strong immediate learning effects, with the representation training showing a 10 to 30 percentage point advantage over the rule training.

But how quickly did participants forget what they had just learned? The American students were tested with new problems one week and then five weeks after training. Consistent with the disheartening observations of many statistics teachers, after 1 week the performance of students who received the rule training had dropped to a median of 30 percent correct solutions, and after 5 weeks it was down to only 20 percent. The performance of the students who received the representation training did not show any decay. Five weeks after training, the median percentage of Bayesian inferences was still 90 percent. These students showed no sign of forgetting what they had learned.

The German students were tested after one week and after three months—a more stringent test of forgetting. After one week, the students who had received the rule training, unlike their American counterparts, showed no sign of forgetting what they had learned. After three months, however, the effect of rule training had eroded to 50 percent, a substantial drop but again a smaller one than in the American study. The most striking result in the German study was this: Three months after training, the performance of the students in the representation training group still showed no sign of flagging. The initial training effect of learning did not only remain stable after three months—it increased to 100 percent correct solutions!

The results of these training studies indicate that when students learn to use proper representations, the problem of memory decay—in this context, forgetting how to think about uncertainties—can be largely overcome. At the same time, when students learned to play with representations—as opposed to mechanically inserting probabilities into equations—the differences between the American and German students also disappeared.

TEACHING REPRESENTATIONS

These results bode well for efforts at teaching statistical thinking. Because the representation training takes only one to two hours including testing, it can be used, for instance, in high school curricula to teach young people how to interpret pregnancy test results and statistics about the hazards of drug use. Similarly, it can be used in medical schools to teach physicians how to estimate the chance of cancer given a positive test and in law schools to teach students how to draw conclusions on the basis of uncertain evidence such as a DNA match. All this can be done by a computerized tutorial or a human teacher. Computerized tutorials attract the attention of all kinds of people, and the participants in our training studies showed a remarkable degree of involvement and motivation to succeed.

These results may be good news for instructors who design college-preparatory curricula that teach young people how to reckon with risk in a technological world and for those heretofore unfortunate souls among us who are charged with teaching undergraduate statistics. The struggle for statistical literacy is more likely to succeed if educators and students arm themselves with information representations suited to the human mind.

Dare to Know

Aristotle once divided our world into two realms, the heavenly world of immutable regularities and certain knowledge and the messy world of change and uncertainty. In Western culture, people wanted to live in the world of certain knowledge, not in a world that is hard to understand and predict and where accidents and errors reign. For centuries, mathematicians believed they lived in the world of absolute certainty, as did theologians and their followers. The Reformation and Counter-Reformation, however, greatly eroded the empire of certainty. During the Inquisition, for instance, torture was seen as a means of discovering definitive truth, and this noble goal was viewed as justifying its ignoble means. It may not be an accident that the mathematical theory of probability emerged only after this religious turmoil—or that as the new, more modest standard of evidence that grew out of probability theory spread, the use of torture began to decline. By the mid–seventeenth century, a new standard of reason had emerged, one that did not aspire to certainty but to reasonable judgments under uncertainty.[27]

In our modern technological societies, Aristotle's two realms still coexist. Most of us prefer to wander back and forth between them without much reflection. In sports, for instance, we revel in the world of uncertainty. We know that the outcomes of a game are a mixture of strategy and accident. In sports, as well as in the stock market and other competitive situations, we enjoy uncertainty; otherwise all the excitement, anticipation, and surprise would be lost. In other aspects of our lives, however, we cherish the illusion of certainty and turn our backs on the uncertainty we love so much in the realm of competition and entertainment. For instance, when it comes to food and health, many people take an authority's or a journalist's opinion as definitive without checking to see if it is reasonable. One goal of this book is to make you, the reader, aware of the illusion of certainty. A second goal is to provide you with tools to help you understand risks and how to communicate these effectively to others. These tools for turning innumeracy into insight, such as replacing relative with absolute risks and probabilities with natural frequencies, are easy to learn.

Many have argued that sound statistical thinking is not easily turned into a "habit of mind."[28] This claim has been used by authorities ranging from politicians to physicians to justify withholding information from the general public. I disagree with this habit-of-mind story. The central lesson of this book is that people's difficulties in thinking about numbers need not be accepted, because they can be overcome. These difficulties are not simply the mind's fault. Often, the solutions can be found in the mind's environment, that is, in the way numerical information is presented. With the aid of intuitively understandable representations, statistical thinking can become a habit of mind.

This book began with Benjamin Franklin's declaration that in this world nothing is certain but death and taxes and H. G. Wells's vision of a world in which citizens master statistical thinking along with reading and writing. I strongly believe that Wells's dream is worthy of our time and effort. At the close of the book, we see that this dream calls for two things: knowing and daring. One without the other is like a pair of scissors with only one blade; both are essential. So let me close with a reprise of Kant's challenge: *Sapere aude*—dare to know.

Absolute risk reduction. A measure of the efficacy of a treatment in terms of the absolute number of people saved. For instance, if a treatment reduces the number of people who die of a disease from 6 to 4 in 1,000, then the absolute risk reduction is 2 in 1,000, or 0.2 percent.

Average. A measure for the central tendency of a sample of observations. The term *average* is most often used for the arithmetic mean, but sometimes also for the median. For instance, suppose the yearly income of five brokers is $80,000, $90,000, $100,000, $130,000, and $600,000. The arithmetic mean is the sum of these values divided by their number, that is, $200,000. The median is obtained by ranking the values (as above) and taking the one in the middle, that is, $100,000. When the distribution is asymmetric, as it is often with income, the mean and the median are not the same, and it can be the case that most people earn less than the mean.

Base rate. The base rate of an attribute (or event) in a population is the proportion of individuals manifesting that attribute (at a certain point in time). A synonym for base rate is *prevalence.* See also *incidence rate.*

Bayes's rule. A procedure for updating the probability of some hypothesis in the light of new evidence. The origin of the rule is attributed to the Reverend Thomas Bayes. For the simple case of a binary hypothesis (H and not-H, such as cancer and not cancer) and data D (such as a positive test), the rule is:

$$p(H|D) = p(H)p(D|H)/[p(H)p(D|H) + p(\text{not-}H)p(D|\text{not-}H)]$$

where $p(D|H)$ is the posterior probability, $p(H)$ is the prior probability, $p(D|H)$ is the probability of D given H, and $p(D|\text{not-}H)$ is the probability of D given not-H.

Many professionals have problems understanding this rule. The interesting point is that the calculation of $p(H|D)$ becomes more intuitive and much simpler when the input is in natural frequencies rather than probabilities. For natural frequencies, the rule is:

$$p(\text{H}|\text{D}) = a/(a + b)$$

where *a* is the number of D and H cases, and *b* is the number of D and not-H cases.

Clouded thinking. A form of innumeracy, in which a person knows about the risks but not how to draw conclusions or inferences from them. For instance, physicians often know the error rates of mammography and the base rate of breast cancer, but not how to infer from this information the chances that a woman with a positive test actually has breast cancer. Mind tools for overcoming clouded thinking, such as natural frequencies, are representations that facilitate drawing conclusions.

Conditional probability. The probability that an event A occurs given event B, usually written as $p(\text{A}|\text{B})$. An example of a conditional probability is the probability of a positive screening mammogram given breast cancer, which is around .90. The probability $p(\text{A})$, for instance, is not a conditional probability. Conditional probabilities are notoriously misunderstood, and that in two different ways. One is to confuse the probability of A *given* B with the probability of A *and* B; the other is to confuse the probability of *A* given *B* with the probability of *B* given *A*. One can reduce this confusion by replacing conditional probabilities with natural frequencies.

Degrees of belief. One of the three major interpretations of probability (besides relative frequencies and propensities). The probability of an event is the subjective degree of belief a person has in that event. Historically, degrees of warranted belief entered probability theory from applications in the courtroom, such as the credibility of witnesses. Degrees of belief are constrained by the laws of probability (for example, probabilities need to add up to 1), that is, beliefs need to follow these laws to qualify as subjective probabilities.

Early detection. Early detection of a disease is the goal of screening for it. Early detection can reduce mortality. Early detection, however, does not imply mortality reduction. For instance, if there is no effective therapy, then early detection, including treatment, will not reduce mortality.

Error. A test can result in one of two errors, a false positive or a false negative. These errors can result from various sources, including human error (for example, the laboratory assistant confuses two samples or labels, or enters the wrong result into the computer) and medical conditions (for example, a positive HIV test can result from rheumato-logical diseases and liver diseases that have nothing to do with HIV). Errors can be reduced but not completely eliminated, and they may even be indispensable to adaptation and survival, as the copying errors (mutations) in DNA illustrate.

Evidence-based medicine. To treat patients consistent with the best scientific evidence available, taking the values of the patient into consideration.

Expert witness. A person identified by the court as an expert allowed to testify before the court to facts, to draw conclusions by bringing together the available data, and to testify to matters not accessible to lay knowledge, such as insanity, testament capacity, and standards of care.

False negative. A test result in which the test is negative (for example, a pregnancy test finds no sign of pregnancy) but the event is actually there (the woman is pregnant); also called a "miss."

False negative rate. The proportion of negative tests among people with the disease or condition. It is typically expressed as a conditional probability or a percentage. For instance, mammography screening has a false negative rate of 5 to 20 percent depending on age, that is, 5 to 20 percent of women *with* breast cancer receive a negative test result. The false negative rate and the sensitivity (hit rate) of a test add up to 100 percent. The false negative rate and the false positive rate are dependent: To decease one is to increase the other.

False positive. A test result in which the test is positive (for example, a positive pregnancy test) but the event is not extant (the woman is not pregnant); also called a "false alarm."

False positive rate. The proportion of positive tests among people without the disease or condition. It is typically expressed as a conditional probability or a percentage. For instance, mammography screening has a false positive rate of 5 to 10 percent depending on age, that is, 5 to 10 percent of women without breast cancer nevertheless receive a positive test result. The false positive rate and the specificity (power) of a test add up to 100 percent. The false positive rate and the false negative rate are dependent: To decrease one is to increase the other.

Franklin's law. "Nothing is certain but death and taxes." A reminder that in all human conduct, uncertainty is prevalent as the result of human and technical errors, limited knowledge, unpredictability, deception, or other causes.

Frequencies. A number of observations in a class of events. Frequencies can be expressed as relative frequencies, absolute frequencies, or natural frequencies.

Guilt probability. The probability p(guilt|evidence) that a person is guilty given the evidence, such as a DNA match.

Ignorance of risks. An elementary form of innumeracy in which a person does not know, not even roughly, how great a relevant risk is. It differs from the illusion of certainty (for example, "smoking cigarettes does not cause lung cancer") in that the person is aware that there are uncertainties, but does not know how great they are.

Illusion of certainty. The belief that an event is absolutely certain although it may not be. For instance, people tend to believe that the results of modern technologies, such as HIV testing, DNA fingerprinting, ordinary fingerprinting techniques, medical tests, or even the mechanical vote-counting machines used in elections are certain, that is, error-free. The illusion can have benefits, such as reassurance, but also costs, such as suicide after a false positive HIV test. With respect to morals, religion, and political values, the illusion of certainty may be a requirement for being accepted by a social group, fostering social control.

Incidence rate. Unlike prevalence (base rate), which refers to the proportion of individuals in a population manifesting an attribute (or event) *at a certain point in time,* the incidence rate is the proportion of individuals in a given population developing this attribute *within a specified time interval.* For instance, the proportion of men with prostate cancer at age 50 is a prevalence; the proportion of men who will develop prostate cancer between 50 and 60 is an incidence rate.

Independence. Two events are independent if knowing the outcome of one does not inform us about the outcome of the other. Formally, two events A and B are independent if the probability p(A&B) that A and B occur together is the product of p(A) times p(B). The concept of independence is crucial, for instance, to evaluating a match between a defendant's DNA and that found on the victim. Assume only 1 out of 1 million men show such a match. If the DNA of all a country's citizens are in a data bank, and one citizen's DNA is randomly selected, then the probability of a match is about 1 in a million. If the defendant, however, has an identical twin, the probability that the twin also shows a match is 1 (except for procedural errors), not 1 in 1 million. Similarly, if the defendant has brothers, the probability that they match is considerably higher than for the general population. The DNA of relatives is not independent; knowing that one matches increases the chances that the relative also matches.

Informed consent. The ideal that the patient should be informed about the pros and cons of a treatment and its alternatives, and on this basis should decide whether he or she wants to enter treatment. Today's medical practice has not yet generally reached this ideal, partly because patients want to be taken care of rather than be informed, and partly because physicians prefer to decide what treatment to apply. The legal doctrine of informed consent deals with the voluntary consent of humans to biomedical research and medical treatment, the question of how much disclosure is enough (an issue in malpractice trials), the competence of the patient (an issue in children and the mentally retarded), and the right to refuse treatment.

Innumeracy. The inability to think with numbers. Statistical innumeracy is the inability to think with numbers that represent uncertainties. Ignorance of risk, miscommunication of risk, and clouded thinking are forms of innumeracy. Like illiteracy, innumeracy is curable. Innumeracy is not simply a mental defect "inside" an unfortunate mind, but is in part produced by inadequate "outside" representations of numbers. Innumeracy can be cured from the outside.

Life expectancy. The expected number of years remaining to be lived by persons of a particular age.

Mind tools. Means, such as Franklin's law and proper representations of risks, used to overcome the illusion of certainty and innumeracy.

Miscommunication of risks. A form of innumeracy, in which a person knows the risks of some event or action but does not know how to communicate these so that others understand them. Mind tools for overcoming miscommunication are representations that facilitate understanding.

Mortality reduction. A measure of the benefit of a treatment in terms of lives saved. The mortality reduction can be represented in many ways, including relative risk reduction, absolute risk reduction, and increased life expectancy.

Natural frequencies. Numbers that correspond to the way humans encountered information before the invention of probability theory. Unlike probabilities and relative frequencies, they are "raw" observations that have not been normalized with respect to the base rates of the event in question. For instance, a physician has observed 100 persons, 10 of whom show a new disease. Of these 10 persons, 8 show a symptom, whereas 4 of the 90 without disease also show the symptom. Breaking these 100 cases down into four numbers (disease and symptom: 8; disease and no symptom: 2; no disease and symptom: 4; no disease and no symptom: 86) results in four natural frequencies 8, 2, 4, and 86. Natural frequencies facilitate Bayesian inferences. For instance, if the physician observes a new person with the symptom, the physician can easily see that the chance that this patient also has the disease is 8/(8 + 4), that is, 2/3. If the physician's observations, however, are transformed into conditional probabilities or relative frequencies (for example, by dividing the natural frequency 4 by the base rate 90, resulting in .044, or 4.4 percent), then the computation of this probability becomes more difficult and requires Bayes's rule for probabilities. Natural frequencies help people to make sound conclusions, whereas conditional probabilities tend to cloud minds.

Negative test result. Typically good news. That is, no sign of a disease has been found.

Number needed to treat (NNT). A measure of the efficacy of a treatment. For instance, if mammogram screening eventually saves the life of 1 in 1,000 participating women, the NNT (to save one life) is 1,000. In other words, 999 women do not benefit in terms of mortality reduction. NNT is also used to measure the harm of a treatment, such as when about 1 in 7,000 women who take oral contraceptives get thromboembolism, the NNT (with oral contraceptives to cause one case of thromboembolism) is 7,000. In other words, 6,999 do not show this side effect.

Number of days/years gained or lost. A measure of the efficacy of a treatment or habit in terms of increase or decrease in life expectancy. For instance, 30 years of smoking one or two packs of cigarettes a day results in an average loss of 2,250 days, or about 6 years of life.

Number of people who match a characteristic. A transparent way to express the meaning of an observed match between the characteristics of a defendant and the evidence. An example is the statement "1 out of

10,000 men in this population shows a match." In contrast, the following single-event statement (a random match probability) is mathematically equivalent but can easily lead to misunderstanding in court: "The probability that this match occurred by chance is 1 in 10,000, or .01 percent."

Odds. The ratio of two probabilities (of the two possible outcomes of an event) is called odds. For instance, the probability of getting a six by throwing a fair die is 1/6, and the probability of not getting a six is 5/6. Thus the odds of getting a six are 1 to 5.

Percentages. There are three kinds of percentages. One is single-event probabilities multiplied by 100 (as in, Washkansky has an 80 percent chance of survival). With this type of statement one can produce the same misunderstanding as with single-event probabilities. A second kind is conditional probabilities multiplied by 100. With this form of communication, one can produce the same confusion as with conditional probabilities. A third kind of percentage is (unconditional) relative frequencies multiplied by 100. An example is the finding of the 1962 Gallup poll that only 38 percent of adult Americans knew that cigarettes caused lung cancer. Such a percentage is easy to comprehend as long as the reference class is clear.

Placebo effect. A placebo operates through the mind instead of the body. For instance, when a physician gives a patient a sugar pill or injection that contains no ingredients known to influence the patient's cold symptoms or rash, but the patient nevertheless experiences relief, that is the placebo effect. *Placebo* is Latin for "I shall please." Placebos do not work all the time or for all diseases; their effect seems to depend on how strongly the patient believes that the treatment would actually be effective. Placebos are a challenge to the ideal of informed consent.

Positive mammogram. The result of a mammography is typically classified as positive (suspicious) or negative. Positive results are usually distinguished into three levels of suspicion, such as "additional evaluation needed," "suspicion of malignancy," and "malignant by radiological criteria." The vast majority of positive screening mammograms (more than 90 percent) fall into the lowest level of suspicion.

Positive predictive value. The proportion of people among all those who test positive who actually do have the disease (or condition): i.e. the true positives divided by the total number who test positive.

Positive test result. Typically not good news. That is, a possible sign of a disease has been found.

Posterior probability. The probability of an event after a diagnostic result, that is, the updated prior probability. It can be calculated from the prior probability using Bayes's rule.

Prevalence. See *base rate.*

Principle of indifference. When the prior probabilities (or base rates) are not known, the principle of indifference can be invoked. In the simplest case with two alternatives, the principle of indifference would assign each alternative a prior probability of one-half, for three alternatives the priors would be one-third each, and so on.

Prior probability. The probability of an event prior to new evidence. Bayes's rule specifies how prior probabilities are updated in the light of new evidence.

Probability. A measure that quantifies the uncertainty associated with an event. If an event A cannot happen, the probability $p(A)$ is zero; if an event happens with certainty, $p(A)$ is 1; otherwise the values of $p(A)$ are between zero and 1. For a set of events, A and B, which are mutually exclusive and exhaustive, the probabilities of the individual events add up to 1.

Proficiency tests. A method to estimate the accuracy of a diagnostic testing procedure, such as DNA fingerprinting. For instance, a number of samples (say, DNA fingerprints) are sent to a large number of laboratories, which then independently analyze the evidence and determine whether some of the samples match. Results can be used to estimate the false negative and false positive rates, as well as the quality of the individual laboratories. Proficiency tests can be blind (that is, the laboratories and technicians do not know that they are being tested) or not blind (they do know), and internal or external (the samples are only analyzed in house or by external laboratories).

Propensities. One of the three major interpretations of probability (besides relative frequencies and degrees of belief). The probability of an event is defined by physical design. Historically, propensities entered probability theory from gambling, such as from the physical design of dice and roulette wheels. Propensities are limited to events whose underlying design or causal mechanism is known.

Prosecutor's fallacy. The confusion of the probability *p*(match) that the defendant matches the characteristics of the evidence available with the probability *p*(not guilty|match) that the defendant is not guilty given that he or she shows a match. Because *p*(match) is typically very small, such as when the evidence is a DNA trace, the confusion "serves" the prosecution because it makes the probability that the defendant is innocent appear equally small.

Randomized trial. A method for estimating the benefits of a treatment that uses randomization as a method of control. Participants in a randomized trial are randomly assigned to either a treatment (for example, prostate cancer screening) or to a control group (no prostate cancer screening). After a certain number of years, the two groups are compared on criteria, such as mortality, to determine whether the treatment has been effective. Randomization allows the control of variables—such as age, education, and health—that could be alternative explanations (besides the treatment) for an observed difference in mortality. The randomized clinical trial uses the same idea of control as in experiments with random assignment to groups.

Random match probability. The relative frequency of a trait, or combination of traits, in a population. That is, the random match probability is the probability that a match occurs between a trait (for example, a DNA pattern) found on the scene of a crime and a person randomly drawn from a population.

Reference class. A class of events or objects to which a probability or frequency refers. In the frequency interpretation of the concept of probability, there is no probability without a specified reference class. This view excludes single-event probabilities, which, by definition, specify no reference class.

Relative frequencies. One of the three major interpretations of probability (the others are degrees of belief and propensities). The probability of an event is defined as its relative frequency in a reference class. Historically, frequencies entered probability theory through mortality tables that provided the basis for calculating life insurance rates. Relative frequencies are constrained to repeated events that can be observed in large numbers.

Relative risk reduction. A measure of the efficacy of a treatment in terms of the relative number of people saved. For instance, if a treatment reduces the number of people who die from 6 to 4 in 1,000, then the relative risk reduction is 33.3 percent. Reporting relative risks is popular because the numbers look larger than the absolute risk reduction (which would be 2 in 1,000, or 0.2 percent). Relative risks do not convey how large, in absolute terms, the risk is, and as a consequence, are often misunderstood. For instance, if a treatment reduces the number of people who die from 6 to 4 in 10,000, the relative risk reduction is still the same (33.3 percent), although the absolute risk reduction has decreased to 0.02 percent.

Reliability. The extent to which a test produces the same results under different conditions (such as repeated measurements). High reliability is necessary but does not guarantee high validity.

Risk. Uncertainty associated with an event that can be quantified on the basis of empirical observations or causal knowledge. Frequencies and probabilities are ways to express risks. Unlike in its everyday use, the term *risk* need not be associated with a harm; it can refer to a positive, neutral, or negative event.

Screening. The testing of a symptomless population in order to detect cases of a disease at an early stage. The term *screening* is also used outside of medicine, for instance, when a population is screened for a DNA profile.

Sensitivity. The percentage of individuals with a disease who test positive in a test, that is, who are correctly classified as having the disease. Formally, the sensitivity is the conditional probability p(positive|disease) of a positive test result given the disease. The sensitivity and the false negative rate add up to 100 percent. The sensitivity is also called the "hit rate."

Sensitivity of mammography. The proportion of women who test positive on mammography among those who have breast cancer. It ranges between about 80 and 95 percent, with the lower values in younger women. The sensitivity of mammography depends primarily on the ability of the radiologist to identify breast cancers and on the rate at which breast cancers double in size between screening examinations.

Single-event probabilities. A probability associated with a singular event for which no reference class is known or specified. For instance, the statement "there is a 30 percent chance that it will rain tomorrow" is a probability statement about a singular event—it either rains or does not rain tomorrow. In contrast, the statement that it will rain on 10 days in May is a frequency statement. The latter statement can be true or false; a single-event probability by itself can never be proven wrong (unless the probability is zero or 1). Single-event probabilities can lead to the miscommunication because people tend to fill in different reference classes. For instance, people understand that the statement "there is a 30 percent chance of rain tomorrow" means that it will rain for 30 percent of the time, or in 30 percent of the area, or on 30 percent of the days that are like tomorrow. This miscommunication can be avoided by using frequencies instead of single-event probabilities because frequencies spell out a reference class.

Source probability. The probability $p(source|match)$ that a person is the source of a trace given a match. (An example for a trace is blood found at the scene of a crime.)

Specificity. The percentage of individuals without a disease who test negative in a test, that is, who are correctly classified as not having the disease. Formally, the specificity is the conditional probability $p(negative|no\ disease)$ of a negative test result given no disease. The specificity and the false positive rate add up to 100 percent. The specificity is also called the "power" of a test.

Specificity of mammography. The proportion of women who test negative on mammography among those who do not have breast cancer. It ranges between 90 and 95 percent, with the lower values for younger women.

Uncertainty. An event or outcome that is not certain but may or may not happen is *uncertain.* When the uncertainty is quantified on the basis of empirical observations, it is called "risk."

Validity. The extent to which a test measures what it was intended to measure. High reliability is necessary but does not guarantee high validity.

NOTES

CHAPTER 1. UNCERTAINTY

1. For details see Chapter 5.
2. Franklin (1789). In a letter, Benjamin Franklin wrote: "Our Constitution is in actual operation; everything appears to promise that it will last; but in this world there is nothing certain but death and taxes."

CHAPTER 2. THE ILLUSION OF CERTAINTY

1. The tendency to see certainty instead of uncertainty, structure instead of noise, and cause and effect instead of accidental contingencies has been described in other contexts by historians and psychologists alike (e.g., Fischhoff, 1982; Thinès, Costall, and Butterworth, 1991).
2. See Gigerenzer and Murray (1987, Chapter 3).
3. Stigler explains how Galton calculated this number (1999, Chapter 6).
4. Fingering fingerprints. (2000, December 16) *The Economist*, 103–104.
5. Dora Heller drew my attention to the possible origins of the alleged dangers of consuming cherries and water at the same time. Fruit that has not been washed is covered with microorganisms that can lead to fermentation processes in the stomach, which in turn can cause pain. Normally, these microorganisms are destroyed by stomach acid. However, if a person drinks a large amount of liquid while eating, the acid is diluted, and is thus hindered from doing its job well, and stomach pain can result. But washing fruit before eating it is very likely to take care of this problem.
6. Daston (1987).
7. Evers, Dworschak, et al. (2000); Evers, Neubacher, et al. (2000).
8. Katz (1984, pp. 166–169).
9. I have omitted details that reveal the identities of the participants, and report only their professional status. I also have rearranged and shortened the discussion to make it easier to follow.
10. Humphrey (2000).
11. Harrington (1997).
12. Kant (1784).

CHAPTER 3. INNUMERACY

1. In the 1990s, the toy maker Mattel came out with a talking Barbie doll that could utter a few phrases, including the two quoted here. Feminist groups objected to the doll's unenlightened statements, which they alleged would encourage girls to be more concerned with shopping than with school or careers. One group bought 500 talking Barbie dolls and 500 talking G.I. Joe dolls just before Christmas, exchanged their voice boxes, and returned them to the store. As a result, some little girls must have unwrapped their talking Barbie on Christmas morning only to hear her say, "Eat lead, Cobra!" (Bondy, 1999).

2. This statement is quoted from *How to Lie with Statistics* (Huff, 1954/1993), where it serves as an epigraph. No reference is given. I have searched through scores of statistical textbooks in which it has since been quoted and found none where a reference was given. I could not find this statement in Wells's work either. Thus, the source of this statement remains uncertain, another example of Franklin's law.

3. The survey was conducted by the Emnid Institute, a polling organization; see *Süddeutsche Zeitung Magazin* (1998, December 31).

4. In the literature, the economist Frank Knight (1921) is credited with making the distinction between risks and uncertainties, and it is almost universally claimed that his distinction is that between uncertainties that can and cannot be quantified. However, Knight in fact proposed a different distinction, that between objective probabilities, such as empirical frequencies ("risk") and subjective probabilities, such as degrees of belief ("uncertainty"). In his own words, "We can also employ the terms 'objective' and 'subjective' probability to designate the risk and uncertainty respectively . . ." (p. 233). In Knight's view, people can assign subjective quantitative probabilities to every conceivable event, but this alone does not qualify these uncertainties as risks (see LeRoy & Singell, 1987). With objective probabilities (that is, risks) Knight meant probabilities everyone would agree on. The distinction I am making in this book is consistent with the one Knight actually made, that is, between uncertain events where we do and do not have empirical data to quantify (at least roughly) the uncertainties involved. The distinction between "uncertainty" and "risk" is, of course, a continuum rather than an opposition, because the amount of empirical evidence available falls on a continuum. Writers such as Mary Douglas (1992) and Ulrich Beck (1986) use the term "risk" in a different way, one that explicitly includes moral and political values. Difficulties in noticing, understanding, and evaluating risks arise not only from the illusion of certainty, or from innumeracy, but from a multitude of emotional, social, and political factors (e.g., Slovic, 1999).

5. Cited in Katz (1984, p. 136).

6. Katz (1984, p. 134).

7. Proctor (1998, 1999).

8. Ross (1999, p. 40).
9. Statistisches Bundesamt (Federal Statistical Office) (2000b, p. 230).
10. Statistisches Bundesamt (Federal Statistical Office) (2000a, p. 282–283).
11. Bourguet (1987).
12. Porter (1986). For more on Hacking's work, see Hacking (1975; 1990).
13. U.S. Department of Transportation (1998, p. 2).
14. See National Research Council (1989).
15. See Skolbekken (1998).
16. Compare Malenka et al. (1993).
17. Hacking (1975).

CHAPTER 4. INSIGHT

1. This term was originally suggested by psychiatrist John Bowlby (see Nesse and Williams, 1995, p. 138). Anthropologists disagree about what conditions were like in the environment of evolutionary adaptedness, such as the amount of time people spent searching for and preparing food, child-rearing customs, divisions of labor, systems of inheritance, and social group size. Some of these conditions may have varied widely over space and time, whereas others did not. The properties of sunlight and the 24-hour night-day cycle are examples of unvarying conditions.
2. On color constancy as an evolutionary adaptation, see Shepard (1987, 1992); on the immune system as a collection of adaptations, see Nesse and Williams (1995); on cognitive heuristics as adaptations see Gigerenzer et al. (1999).
3. I have changed his name to protect his identity.
4. The information corresponds to the results of the first screening mammography of 26,000 American women over 30 years old (Kerlikowske et al., 1996a, 1996b). During the first screening, in about 0.8 percent of these women, breast cancer was identified. Among women aged 40 to 49 with breast cancer, 87 percent had a positive mammogram, and among women without breast cancer, 7 percent also had a positive mammogram. The result "positive" meant one of three conditions: Most of the positive results were reported as "additional evaluation needed" (93 percent); the others were either "suspicious for malignancy" (5 percent) or "malignant" by radiological criteria (2 percent). Note that the base rate of cancer can vary with age, risk group, and country.
5. The values used in this study differed slightly from those shown in Figure 4-2. In this study, the base rate, sensitivity, and false positive rate were .01, .8, and .1, respectively. The probability of breast cancer given a positive mammogram was .075, or expressed in frequencies, out of 100 persons with a positive mammogram, 7 or 8 actually had breast cancer. The details of this study can be found in Gigerenzer (1996a) and Hoffrage and Gigerenzer (1998).

6. David Eddy's informal study is reported in Eddy (1982).
7. Bayes's rule with probabilities is usually provided in today's social science text-books, whereas the version with natural frequencies corresponds to Thomas Bayes's (1763) original Proposition 5 (see Earman, 1992).
8. As Franklin's law admonishes, however, we would be wise not to take this attribution of origin too literally. As with most rules and laws, it is uncertain who actually discovered Bayes's rule. According to Stigler's law of eponymy, no scientific discovery is named after its actual discoverer. Examples include the Pythagorean theorem, Pascal's triangle, and the Gaussian distribution. It remains to be discovered who actually discovered Stigler's law. In a gripping historical detective story, Stigler (1983) concluded that the odds are 3 to 1 that Nicholas Saunderson rather than Thomas Bayes discovered Bayes's rule. Despite being totally blind from the age of 1, Saunderson mastered all of mathematics and natural philosophy very early in life. At 29, he held the most prestigious academic chair in England, the Lucasian chair of mathematics at Cambridge, which Newton had held before him. He died of scurvy, in 1739, at the age of 56. Bayes, of course, cannot be accused of behaving in accord with the impolite but false interpretation of Stigler's law, namely, "Every scientific discovery is named after the last individual too ungenerous to give due credit to his predecessors" (Stigler, 1980). Bayes never published his treatise, for which the eminent statistician R. A. Fisher (1935) congratulated him. (Fisher opposed the use of Bayes's rule in the sciences; see Gigerenzer et al., 1989, Chapter 3.) It was published posthumously by Richard Price in 1763.
9. Kleiter (1994); Gigerenzer and Hoffrage (1995, 1999).
10. Why do I refer to *natural frequencies* as such rather than simply to *frequencies*? Because there are other frequencies that are normalized just like probabilities. For instance, consider the following version of the mammography problem, which uses such "unnatural" frequencies:

> Eight out of every 1,000 women have breast cancer. Of 1,000 women with breast cancer, 900 will have a positive mammogram. Of 1,000 women without breast cancer, 70 will still have a positive mammogram. Imagine a sample of women who have positive mammograms in screening. How many of these women actually have breast cancer?

This representation causes as much clouded thinking as probabilities do, and for the same reasons (Gigerenzer and Hoffrage, 1999; Lewis and Keren, 1999). It does not correspond to the direct experience illustrated by the tree in Figure 4-2. A physician who has seen 1,000 patients will have seen 8 with breast cancer (Figure 4-2), not 1,000 as above. Of these 8 with cancer, 7 have a positive mammogram. However, in the above version, the natural frequency 7 of 8 is normalized to 900 out of 1,000 (rounded), which takes the information about the base rate (8 cases with cancer) out—just as .9, or 90 percent, does. These frequencies are not natural frequencies because they do not carry information about the base rates. Similarly, it is conditional probabilities—not all probabilities—that tend to cloud thinking. For instance, probabilities and percentages that express simple frequencies (that is, unconditional frequencies), such as "1 percent of women have breast cancer" are easily understood. In contrast, the conditional probability that a woman has breast cancer given that she tests positive is apt to be confused with the conditional probability that a woman tests positive given that she has breast cancer.

11. See Real (1991); Wynn (1998, p. 114); Dehaene (1997, p. 18).

12. See, for example, Barsalou and Ross (1986), Butterworth (1999), Hasher and Zacks (1984), Jonides and Jones (1992), and Sedlmeier, Hertwig and Gigerenzer (1998). Estes (1976) and, early in his career, Brunswik (e.g., 1937) proposed that probability learning arises from frequency learning. Their studies focused on estimations of the frequency of letters, words, and other objects. The question of which events (out of all possible events) are automatically counted is addressed in Brase, Cosmides, and Tooby (1998).

13. Dehaene (1997, p. 50); Antell and Keating (1983).

14. For an overview, see Dehaene (1997) and Gallistel and Gelman (1992).

15. See Dehaene (1997, p. 87).

16. The emergence of mathematical probability—and a new conception of reasonableness that acknowledges uncertainty rather seeking certainty—is described in Daston (1988) and Hacking (1975). For more on the "probabilistic revolution," see Krüger, Daston, and Heidelberger (1987) and Krüger, Gigerenzer, and Morgan (1987). On how notions of chance, randomness, and probability have changed the sciences and everyday life, see Gigerenzer et al. (1989).

17. Miller, Smith, Zhu, and Zhang (1995).

18. Feynman (1967, p. 53). On the distinction between an algorithm and a representation, see Marr (1982).

CHAPTER 5. BREAST CANCER SCREENING

1. Roberts (1989, pp. 1153–1154).

2. Proctor (1999). Wilhelm Conrad Röntgen, who discovered the X ray in 1895, called these new rays "X rays" because he did not know what they were made of.

3. In this section, I draw on Kerlikowske (2000).

4. Cockburn (1995).

5. Kerlikowske (2000).

6. These trials included some 280,000 women (Nyström et al., 1996) and provided the raw data so that the various representations of benefits could be calculated (I follow Mühlhauser and Höldke, 1999, here).

7. To simplify the explanation, I have rounded the numbers up (4 and 3) rather than using the actual raw numbers of the Swedish trials, which were 3.6 and 2.9. The actual numbers give slightly smaller values for the benefits of mammography screening: a relative risk reduction of about 20 percent, an absolute risk reduction of 7 in 10,000, and a number needed to treat of 1,429. An important result in these trials is that the total mortality—from breast cancer and other causes—was the same in the screening group as in the control group. That is, screening seems to reduce the number of women who die from breast cancer, but not the number of women who die.

8. Salzmann, Kerlikowske, and Phillips (1997). These authors found an average additional increase in life expectancy of 2.5 days per woman when they included 40- to 49-year old women who underwent screening every 18 months.

9. Schmidt (1994, p. 69).

10. Kerlikowske (2000). After 10 to 14 years there was, however, a trend toward a mortality reduction. Does this mean that mammography screening of women in their 40s does not immediately result in a mortality reduction, but rather only after ten or more years? The answer seems to be no, because almost all of this reduction occurred in those women who had breast cancer detected at age 50 or older. In other words, if these women had started mammography screening at age 50 years or older (rather than already in their 40s), they would have had the same benefit (Kerlikowske, 1997, p. 83).

11. Note that all benefits are expressed in terms of mortality reduction rather than survival statistics. Survival statistics can be misleading. For instance, consider the general result that there is no mortality reduction with mammography screening in women in their 40s. A woman with invasive breast cancer will die at the same age, say at age 55, whether she had participated in mammography screening and her cancer was detected at age 43, or she had not participated and her cancer was detected at age 45, when she found a lump in her breast. With screening, she would have survived 12 years after the cancer was detected; without screening, she would survived 10 years after detection. This example clarifies that the reporting of survival statistics may be misleading and make screening look useful when it is not. This woman only seemed to live longer with a cancer diagnosis, because the diagnosis was earlier (Kerlikowske, 1997).

12. Kerlikowske (1997, 2000). This conclusion that mammography screening reduces mortality in women age 50 or higher, however, has been debated. Gøtzsche and Olsen (2000; see also Olsen and Gøtzsche, 2001) argue that there have been only two adequately randomized trials for women aged 50 to 69. In the other trials, they argue, there were differences between women in the screening and control groups before the studies began that could have confounded the results (for instance, in the Edinburgh trial the screening group included twice as many women

from the highest socioeconomic stratum as did the control group, and the New York trial excluded more women who already had cancer from the screening group than from the control group). What the authors deem to be the only two adequately randomized trials found no effect of screening on breast cancer mortality (a relative risk reduction of 0.8 percent, which is negligible) and an (unexplained) increase in the total number of deaths in the screening group. The authors conclude that "screening for breast cancer with mammography is unjustified" (p. 129). Similarly, Sjönell and Ståhle (2000) conclude—based on their analysis of data from some 2 million mammograms performed in daily clinical practice (as opposed to in randomized trials)—that screening leads to no significant reduction in breast cancer mortality in women aged 50 to 69.

13. Kerlikowske (2000, pp. 897, 901).
14. Kerlikowske (2000, p. 902); see Lerman et al. (1991).
15. Kerlikowske (1997, 2000).
16. NIH Consensus Statement (1997).
17. Metsch et al. (1998).
18. Mühlhauser and Höldke (1999, pp. 105–106).
19. See Elmore et al. (1998); Lerman et al. (1991).
20. The University of California San Francisco Mobile Mammography Screening program (Kerlikowske et al., 1996a; 1996b; see also Mühlhauser and Höldke, 1999, pp. 103–104). This "1 out of 10" figure is averaged across age groups. The errors reported can vary considerably depending on how the presence of cancer is measured. For instance, one meta-analysis found false positive rates as low as 1 percent to 7 percent over a one-year screening interval but as high as 29 percent over a two-year interval (Mushlin, Kouides, and Shapiro, 1998). The proportion of women with cancer among those who test positive during routine screening also depends on the specific type of "positive" diagnosis. In the University of California San Francisco study, most positive mammograms were labeled as needing "additional evaluation" (93 percent), and the remainder as "suspicious for malignancy" or "malignant" by radiological criteria. Among the latter two types of cases, about 1 in 2 women who tested positive actually had breast cancer.
21. See Elmore et al. (1998); Kerlikowske et al. (1996a; 1996b).
22. Elmore et al. (1998). The results are based on the records of Harvard Pilgrim Health Care, a health maintenance organization that serves nearly 300,000 adults in and around Boston.
23. Koubenec (2000).
24. Lerman et al. (1991, p. 259).
25. Ernster and Barclay (1997).
26. Kerlikowske (2000, p. 900); Mühlhauser and Höldke (1999).
27. Schwartz et al. (2000).

28. Proctor (1999, p. 83).
29. National Academy of Sciences Committee (1990) on the biological effects of ionizing radiations (see also Jung, 1998; Mühlhauser and Höldke, 1999, pp. 106–107). Schmidt (1990) estimated the proportion of breast cancer cases induced by radiation to be 1 percent.
30. Jung (1998, p. 341); Mühlhauser and Höldke (1999, p. 106). The absolute risk in Figure 5-1 refers to a dose of radiation roughly corresponding to that emitted during two screening mammograms. The dose is for two screening sessions, each including four exposures, that is, two exposures of each breast.
31. Kerlikowske (2000). The costs of mammograms vary considerably. For instance, in Germany, the cost of a mammogram ranges between about $20 if it is covered by health insurance and $110 if it is not covered (as with screening).
32. Elmore et al. (1998).
33. Domenighetti (2000).
34. Dolan, Lee, and McDermott (1997).
35. Woloshin et al. (2000).
36. Schwartz et al. (2000).
37. Black, Nease, and Tosteson (1995, p. 730). The italicized words were bold in the original source.
38. Black, Nease, and Tosteson (1995). Even by optimistic estimates, that is, if one were only to look at those studies suggesting that mammography reduces mortality in this age group, the absolute risk reduction would be 0.4 (not 60) women in 1,000.
39. Schwartz et al. (2000).
40. Metsch et al. (1998).
41. Slaytor and Ward (1998).
42. Baines (1992).
43. Schindele and Stollorz (2000).
44. Phillips, Glendon, and Knight (1999, p. 142).
45. Black, Nease, and Tosteson (1995, p. 730). The reported average was the median estimate.
46. Admittedly, more than for other cancers, the risk of lung cancer can be strongly affected by behavioral change: An estimated 80 to 90 percent of deaths could be prevented if Americans stopped smoking cigarettes (Cresanta, 1992, p. 439).
47. Hanks and Scardino (1996); Garnick (1994); Wingo et al. (1998).
48. Breast cancer (cover story) (1991). *Time, 137,* 42–49; Baines (1992).
49. Lantz and Booth (1998).
50. Lantz and Booth (1998).
51. See Lantz and Booth (1998).
52. Ellis (1994).

53. Dolan, Lee, and McDermott (1997).
54. Harris et al. (1991).
55. Dawes (1986; 2001).
56. Think of 1,000 women. Thirty-six of them have breast cancer, and 33 of these thirty-six (92 percent) are in the "high-risk" group. Thus, 33 out of 570 high-risk women will develop breast cancer, which is about 1 in 17. For information on how the risk of developing breast cancer depends on age, see Marshall (1993).
57. Hartmann et al. (1999); Hamm, Lawler, and Scheid (1999).

CHAPTER 6. (UN)INFORMED CONSENT

1. Matthews (1995; p. 19); see also Coleman (1987), on the role of statistics in the therapeutic trial in nineteenth-century medicine.
2. Bernard (1865/1957, p. 137).
3. See Matthews (1995, p. 141).
4. Gawande (1999).
5. Bursztajn et al. (1981, pp. 3–19).
6. German-speaking readers may wish to look up www.cochrane.de
7. Pomata (1998, p. xvi).
8. Katz (1984, p. 46).
9. Katz (1984, p. 61).
10. Katz (1984, pp. 90–100).
11. Reagan and Novak (1989, cited in Eddy, 1996, p. 13).
12. I rely here on Katz (1984, pp. 175–206).
13. See Gawande (1999).
14. Marshall (1996, p. 173).
15. Hamm and Smith (1998).
16. Kalet (1994).
17. Center for the Evaluative Clinical Sciences Staff (1996).
18. Eddy (1996, pp. 5, 319).
19. See Center for the Evaluative Clinical Sciences Staff (1996, p. 135).
20. Chang (2000).
21. The U.S. Preventive Services Task Force Staff (1996, p. 123) writes in its report: "Even if the need for treatment is accepted, the effectiveness of available treatments is unproven. Stage C [local extra-capsular penetration] and Stage D [metastatic] disease are often incurable, and the efficacy of treatment for Stage B [palpable, organ-confined cancer] is uncertain."
22. Hanks and Scardino (1996).
23. The recommendation of the U.S. Preventive Services Task Force Staff (1996, p. 119) reads as follows: "Routine screening for prostate cancer with digital rectal ex-

amination, serum tumor markers (e.g., prostate-specific antigen), or transrectal ultrasound is not recommended." The Canadian Task Force on the Periodic Health Examination (CTF) also recommends against the routine use of PSA. It concluded that the evidence was not sufficient to recommend that physicians discontinue the use of digital rectal exams in men aged 50 to 70. The American Cancer Society, in contrast, recommends annual screening with PSA starting at age 50, and annual screening with digital rectal exam starting at age 40.

24. This includes radical prostatectomy, radiation therapy, and hormone treatment; see U. S. Preventive Services Task Force Staff (1996). Although not a principal argument against prostate cancer screening, the economic implications are interesting. The first year of mass screening of all American men over the age of 50 would cost the country $12 billion to $28 billion.

25. Gigerenzer (1996a); Hoffrage and Gigerenzer (1998).

26. Mandel et al. (1993).

27. Metsch et al. (1998).

28. This strategy is also known as "delta R" and has been often proposed as the correct strategy for estimating the degree of covariation between two dichotomous variables, such as cause and effect or disease and symptom (McKenzie, 1994).

29. Independent evidence from experiments with laypeople confirms that there is little intra-individual consistency in diagnostic inferences with probabilities (Gigerenzer and Hoffrage, 1995).

30. Eddy (1982).

31. Berner (1997).

32. His name has been changed.

33. See also Koubenec (2000).

CHAPTER 7. AIDS COUNSELING

1. The case is reported in Stine (1999, p. 359).

2. Busch (1994); Haley and Reed (1994).

3. For more information on the issues discussed in this and the following paragraphs, see Stine (1999).

4. Stine (1999, p. 126).

5. See Stine (1999, p. 31). The following cases are also reported in Stine, pp. 399–405.

6. Stein et al. (1998).

7. Cited in Stine (1999, p. 413).

8. Altman (2000).

9. Stine (1996, pp. 333, 338).

10. Stine (1999, p. 350).

11. Stine (1999, p. 358).

12. Bundesamt für Gesundheit (BAG) in collaboration with the Eidgenössische Kommission für Aids-Fragen (EKAF) (2000).

13. Månsson (1990).

14. Stine (1999, p. 388).

15. Stine (1999, pp. 358, 389).

16. All studies agree that false positives occur, but how often is less clear. This uncertainty has several reasons, including: What constitutes a positive Western blot test has not been standardized (various agencies use different reagents, testing methods, and interpretation criteria; see Stine, 1996, p. 335); the results of the repeated ELISAs and the Western blot test are not independent (that is, one cannot simply multiply the individual false positive rates of the tests to calculate the combined false positive rate; Spielberg et al., 1989); and the higher the prevalence in a group, the lower the specificity seems to be for this group (Wittkowski, 1989). For instance, the German Red Cross achieved a combined (that is, ELISAs and Western blot) specificity for first-time blood donors of 99.98 percent (Wittkowski, 1989). This corresponds to a false positive rate of 2 in 10,000. From a review of the available data, Gigerenzer, Hoffrage, and Ebert (1998) estimated that the false positive rate is around 1 in 10,000. We also reviewed estimates of false positive rates, sensitivity, and prevalence.

    The two principal methods of estimating the sensitivity and specificity of these tests are screening and blind proficiency testing. In screening tests, large numbers of serum samples, for instance, from blood donors, are tested. Blind proficiency testing more closely resembles an experiment than does screening. Samples with and without HIV antibodies are sent to laboratories for analysis (the laboratories are not informed that they are taking part in a study). Details and problems with determining the sensitivity and specificity are reviewed in Schochetman and George (1994).

17. For the calculation of these estimates see Gigerenzer, Hoffrage, and Ebert (1998).

18. This study is described in detail in Gigerenzer, Hoffrage, and Ebert (1998).

19. Enquete Committee of the German Bundestag (1990); Ward (1994).

20. Stine (1999, p. 177).

21. The report of the Enquete Committee of the German Bundestag (1990) estimates the positive predictive value to be less than 50 percent. See Gigerenzer, Hoffrage, and Ebert (1998).

22. See Chapter 14.

23. Bundeszentrale für gesundheitliche Aufklärung (1988–1993).

24. Bundeszentrale für gesundheitliche Aufklärung (2000).

25. Bundeszentrale für gesundheitliche Aufklärung (1993).

26. The analysis was performed by one of my students at the University of Chicago, Ken Greif, in 1994.

27. Reported in Catalan and Pugh (1995).
28. Stine (1999, pp. 378–380).

CHAPTER 8. WIFE BATTERING

1. Dershowitz (1996, p. 101).
2. Dershowitz (1996, p. 105, italics in original).
3. Dershowitz (1996, p. 104).
4. Dershowitz (1996, pp. 101, 108).
5. I. J. Good (1995, 1996), author of *Good Reasoning* and emeritus professor of statistics at the Virginia Polytechnic Institute, seems to have been the first to point out the possible confusion between the probability that a husband will murder his wife given that he has battered her and the probability that a husband has murdered his wife given that he has battered her and that she was murdered by someone. I agree with Good's argument, but am concerned about the way in which he represented the uncertainties. He chose probabilities and odds, as is customary in his profession, rather than natural frequencies, which would have made it easier for readers to understand his compelling argument. To demonstrate the difference, I summarize Good's argument as he made it in the six equations in his 1995 letter to *Nature,* using the corrected figures from his 1996 *Nature* letter. If you have difficulties following the argument in probabilities and odds, you are proving the very point I wish to make. So don't be discouraged; the argument is presented in natural frequencies in the text, where it should be straightforward enough to understand.

Good's argument was that the relevant probability in the Simpson case is not $p(\text{G}|\text{Bat})$ but $p(\text{G}|\text{Bat and M})$, where G stands for "the husband is guilty" (that is, committed the murder in 1994), "Bat" means that "the husband battered his wife," and "M" means that "the wife was actually murdered by somebody in 1994."

Good based his calculations of $p(\text{G}|\text{Bat and M})$ on the odds version of Bayes's rule:

$$\text{posterior odds} = \text{prior odds} \times \text{likelihood ratio,}$$

which in the present case is

$$\frac{p(\text{G}|\text{Bat and M})}{p(\overline{\text{G}}|\text{Bat and M})} = \frac{p(\text{G}|\text{Bat})}{p(\overline{\text{G}}|\text{Bat})} \; \frac{p(\text{M}|\text{G and Bat})}{p(\text{M}|\overline{\text{G}} \text{ and Bat})}$$

where $\overline{\text{G}}$ stands for "the husband is not guilty."

Using the following six equations, marked Good-1 to Good-6, Good explained how to estimate $p(G|\text{Bat and M})$. According to Dershowitz, the probability $p(G|\text{Bat})$ is 1 in 2,500 (Good used a slightly different estimate, 1 in 2,000, but this makes little difference to the result):

$$p(G|\text{Bat}) = 1/2,500. \tag{Good-1}$$

Therefore, the prior odds (O) are

$$O(G|\text{Bat}) = 1/2,499 \approx 1/2,500. \tag{Good-2}$$

Furthermore, the probability of a woman being murdered given that her husband has murdered her (whether he is a batterer or not) is unity, that is,

$$p(M|G \text{ and Bat}) = p(M|G) = 1. \tag{Good-3}$$

Because in the United States there are about 25,000 murders per year, a quarter of these women, and a population of about 250,000,000, Good estimated the probability of a woman being murdered by someone other than her husband as

$$p(M|\overline{G} \text{ and Bat}) = p(M|\overline{G}) \approx 1/20,000. \tag{Good-4}$$

From Equations Good-3 and Good-4 it follows that the likelihood ratio is about 20,000/1; therefore the posterior odds can be calculated thus:

$$O(G|\text{Bat and M}) = 20,000/2,500 = 8. \tag{Good-5}$$

That is, the probability that a murdered, battered wife was killed by her husband is

$$p(G|\text{Bat and M}) = 8/9. \tag{Good-6}$$

Good added that he sent a copy of this note to both Dershowitz and the Los Angeles Police Department and urged that Bayesian reasoning should be taught at the precollege level. I believe that Good's persuasive argument could be understood more easily by experts and ordinary people alike if the information were presented in natural frequencies, rather than probabilities and odds as in the six equations above.

6. Approximately one-quarter of the 25,000 people murdered in the United States each year are female. For a population of about 250 million, half of whom are women, this means that about 6,250 women (one-quarter of 25,000) are murdered out of about 125,000,000 women each year, which is 5 in 100,000. More precise estimates can be made by adjusting for variables such as the woman's age.

7. Dershowitz (1983, pp. xvi, xvii). On the relationship between baseball and statistics see Gigerenzer et al. (1989, Chapter 7).

8. See Ptacek (1999, pp. 4, 74).

9. The following discussion of facts and possible causes of spousal homicide is based in part on Daly and Wilson (1988).

10. Reiss and Roth (1993, p. 80).

11. For this and the following results see Ptacek (1999, pp. 8, 24). Sherman (1992) reports the results of randomized trials conducted by the police to discover what action—such as arrest and a night in jail for the man or simply verbal counseling for the couple—is most effective at reducing the chance of an assailant beating a victim after the police have left the scene.

12. See Ptacek (1999, p. 32).

13. Reiss (1993, p. 69).

14. This figure is the same for white and African American women. See the report of the National Institute of Justice (Tjaden and Thoennes, 2000, and Koss, Koss, and Woodruff, 1991).

15. See Horne (1999).

16. Violence against women increases and decreases according to changes in social and political contexts. Before communism, Russian culture had, like many other cultures, a long history and folklore in which women were believed to possess evil magical powers and to be sinful, which called for rules and punishments to control them. For instance, the Russian wedding custom in which the bride's father passes a whip to her husband seems to have lasted until the late nineteenth century. Atkinson (1977; cited in Horne, 1999).

17. Oberlies (1997, p. 135).

18. Fawcett et al. (1999).

19. McWhirter (1999).

20. Geary (2000, p. 65); A delicate question. (1999, 24 April). *The Economist,* 26.

CHAPTER 9. EXPERTS ON TRIAL

1. *People v. Collins,* 68 Cal.2d 319, *325; Koehler (1997).

2. The reader may wonder why "dark clothing" did not appear in the chart, despite the fact that both the victim and the witness reported this characteristic. The omission by the prosecution was not accidental; evidence had been introduced by the defense that Janet Collins wore light-colored clothing on the day of the crime. For details, see *People v. Collins,* 68 Cal.2d 319 and Koehler (1992).

3. *People v. Collins,* 68 Cal.2d 319, *325.

4. Thompson and Schumann (1987); Balding and Donnelly (1994).

5. Thompson and Schumann (1987).

6. In the case of the AIDS counselors, the confusion is between two conditional

probabilities. However, because $p(\text{match})$ is practically identical with $p(\text{match}|\text{not guilty})$, one can also rewrite the prosecutor's fallacy as the confusion of two conditional probabilities. That is, the prosecutor's fallacy corresponds to "$p(\text{not guilty}|\text{match})$ is the same as $p(\text{match}|\text{not guilty})$."

7. This is an expected value and does not mean that there are exactly two matches in the population. There is some probability that there are zero, one, three, four, or more matches in the population.

8. *People v. Collins*, 68 Cal.2d 319, *331.

9. Gigerenzer et al. (1989, Chapter 7).

10. Gigerenzer (1998); Schrage (1980).

11. Both the probability and the frequency calculations assume that all people in the population of suspects have traces of other people's blood somewhere on their person or property (not necessarily on their boots). Such assumptions affect the size of the population, but do not differentially affect the probability and frequency calculations.

12. Tribe, 1971 (cited in Koehler, 1997, p. 223).

13. Faigman (1999).

14. The term is from Faigman (1999, p. 198), who outlined 12 insights towards recovering from innumeracy.

CHAPTER 10. DNA FINGERPRINTING

1. Later, however, the success of the police was overshadowed by one small discovery. After the police had caught the murderer, they found out that he had been convicted in 1990 of raping his 17-year-old sister, but was released early on probation on the basis of an expert's testimony that he would pose no further danger. However, the police had not entered his conviction into their computer database. If they had, the actual perpetrator would have been a suspect in the first place, and the $2 million screening might have been unnecessary.

2. Stigler (1986); Porter (1986).

3. See Jasanoff and Lynch (1998).

4. Berry (1991).

5. Faigman (1999, p. 85).

6. Boyer (2000).

7. Balding and Donnelly (1994).

8. *R. v. Adams* [1996] 2 Cr. App. R. 467; see Redmayne (2001, p. 58).

9. National Research Council (1996, p. 8).

10. Thompson (1993, p. 57).

11. Hicks (1993, p. 55).

12. Lempert (1991) reviews these and other cases of false positives in DNA matches; see also Koehler (1993b).

13. Koehler, Chia, and Lindsey (1995).

14. Koehler (1997).

15. *State v. Bethune* (cited in Koehler, 1993b, p. 23).

16. For a collection of quotations from legal transcripts see Koehler (1993a; 1993b; 1996).

17. *Frankfurter Allgemeine Zeitung,* 14 April 1998, No. 86, p. 13.

18. *State v. Glover,* Texas, 1992, see Koehler (1993b, p. 30).

19. See Koehler (1993b, p. 28), for a collection of source probability errors in court.

20. Balding and Donnelly (1994, p. 285).

21. Koehler, Chia, and Lindsey (1995).

22. Koehler (1993a, p. 229).

23. Lindsey, Hertwig, and Gigerenzer (2001); Hoffrage et al. (2000); Krauss and Hertwig (2000).

24. Note that in the DNA literature, the term "frequentist" is often applied to frequencies that are not natural frequencies; instead, like single-event probabilities, they refer to a particular person or to a randomly drawn person. For instance, the law professor Richard Lempert describes the reporting of the match probability in the courts thusly: "In the United States, experts typically give this probability in frequentist terms; for example, 'there is one chance in fifty thousand that a randomly selected Caucasian male would have the same DNA profile as that found in both the evidence sample and the sample taken from the defendant.'" (Lempert, 1991, pp. 305–306). Using natural frequencies, in contrast, the expert would say, "Out of every fifty thousand Caucasian males, one would have the same DNA profile." Unlike single-event statements, a statement in terms of natural frequencies directs one's thoughts to the question: How many males can possibly have committed the crime? If we consider a city with 500,000 male inhabitants, we can see that a DNA match by itself does not provide evidence beyond reasonable doubt: we can expect that 10 men will have this profile.

25. Collins and Macleod (1991, p. 210).

26. Collins and Macleod (1991, p. 215).

27. Berry (1991). Jonathan J. Koehler (personal communication) described a Texas case (*State v. Griffith*) in which the judge ruled that it was admissible to use a 50 percent prior probability to generate posterior probabilities of paternity, even after Koehler had testified that to do so would mean that there were hundreds of people who could be deemed more than 99 percent likely to be the father of a particular child. An appellate court upheld this judge's ruling.

28. Daston (1981); Gigerenzer et al. (1989, Chapter 1).

29. A delicate question. (1999, 24 April). *The Economist,* 26.
30. This and the following examples are borrowed from Balding and Donnelly (1994). On the problem of blood relatives and other sources of uncertainty, see Lempert (1991).
31. See Berry (1991, p. 178) and Collins and Macleod (1991, p. 213).
32. Redmayne (1998).
33. In 1998, Monica Lewinsky turned over to investigators a navy blue dress stained with semen. DNA analysis revealed that the source of the semen was President Bill Clinton.
34. Chivers (2000).
35. [1997] 1 Cr.App.R. 369; cited in Redmayne (1998, p. 451) and Redmayne (2001, pp. 71–72).
36. Koehler (1996, p. 877); see also Redmayne (1998, p. 451).

CHAPTER 11. VIOLENT PEOPLE

1. See Monahan (1981). This number refers to the cases in which a psychiatrist predicts future violent acts; that is, in only 1 out of 3 cases in which psychiatrists predict violent acts does one actually occur. The second error—to wrongly predict that there will be no violent act—seems to occur in about 1 out of 10 cases (John Monahan, personal communication, 2000). The use of actuarial tools in risk assessment (see Monahan et al., 2001) may help to improve predictive accuracy.
2. See Grisso and Tomkins (1996, p. 928).
3. Faigman (1999, p. 111).
4. See, for example, Monahan and Wexler (1978). On communicating violence risk, see also the work of Kirk Heilbrun (e.g., Heilbrun et al., 2000).
5. Slovic, Monahan, and MacGregor (2000).
6. The same effect of reference classes has been demonstrated experimentally by Gigerenzer, Hoffrage, and Kleinbölting (1991). I would like to note that Slovic et al.'s explanation for their fascinating results differs from mine: they try to explain the discrepancy by the "frightening images evoked by the frequency format" (p. 290), that is, by images of harmful attacks. However, frequency judgments need not evoke more frightening images than single-event judgments. The Prozac story (Chapter 1) illustrates such a case where a single-event statement evokes more anxiety. The question of frightening images depends, I believe, on the reference class (e.g., psychiatrist's patients versus patient's sexual encounters) specified in the frequency judgment, not on frequency judgments per se. Frequency judgments can refer to any reference class.
7. Slovic and Monahan (1995).

8. Slovic, Monahan, and MacGregor (2000). This result was replicated in a second study in Slovic and Monahan (1995). Mazur and Merz (1994) report the same effect of response scales in a medical judgment context, where the effect was almost as large among experienced clinicians as with clinically naïve persons.

9. See Dawes (1994, pp. 89–90); for tools to improve accuracy see Monahan et al. (2001).

10. Similar effects are known from psychophysics, see Parducci (1965).

11. Schwarz, Hippler, Deutsch, and Strack (1985).

12. See Schwarz (1999); Schwarz and Hippler (1987).

13. See Hansen (1996) and Spellman (1996).

CHAPTER 12. HOW INNUMERACY CAN BE EXPLOITED

1. Huff (1954/1993).

2. Twain (1924, p. 246).

3. Fahey, Griffiths, and Peters (1995).

4. The physicians shall remain unnamed. I am grateful to Dr. Ingrid Mühlhauser for drawing my attention to this leaflet. Another method to influence patient's choice of treatment is called "framing," that is, communicating an outcome in terms of either the probability of dying or that of surviving. For instance, McNeil et al. (1982) report that the preference for a therapy increased when its outcome was framed in terms of the probability of surviving rather than dying. The size of framing effects, howerver, is not very stable (Kühberger, 1998).

5. See Jain, McQuay, and Moore (1998). Desogestrel and gestodene are the main ingredients of many oral contraceptives.

6. Jain, McQuay, and Moore (1998).

7. These numbers are themselves uncertain estimates; for instance, the risk of thromboembolism for a woman on the pill ranges from 2 in 7,000 to 2 in 100,000 (Jain, McQuay, and Moore, 1998).

8. Siegrist (1997).

9. The perils of percentages. (1998, April 18). *The Economist,* 84.

CHAPTER 13. FUN PROBLEMS

1. Laplace (1814/1951). On Laplace, see Gillispie (1997). The rule of succession can be derived from Bayes's rule if one assumes that the prior probabilities are equal, as in the case in which the two hypotheses "the sun will rise every morning" and "the sun will not rise every morning" are initially assigned equal probabilities. To illustrate, I show how the probability of 2/3 (after one observed sunrise) can be obtained from the odds version of Bayes's rule (see Chapter 8, footnote 5):

$$\frac{p(\text{H}|\text{D})}{p(\text{not-H}|\text{D})} = \frac{p(\text{H})}{p(\text{not-H})} \frac{p(\text{D}|\text{H})}{p(\text{D}|\text{not-H})}$$

Here, H (for "hypothesis") stands for "the sun will rise every morning" and not-H for "the sun will not rise every morning," and D (data) for the observation that the sun rose once (only one observation). Assuming uniform priors, $p(\text{H}) = p(\text{not-H}) = 1/2$, we get

$$\frac{p(\text{H}|\text{D})}{p(\text{not-H}|\text{D})} = \frac{p(\text{D}|\text{H})}{p(\text{D}|\text{not-H})}$$

The probability of observing the first sunrise, if H is true, is 1, that is, $p(\text{D}|\text{H}) = 1$. The probability of the same observation, if H is not true, is unknown. In order to get the result of the law of succession, one has to make another assumption of equal probabilities: $p(\text{D}|\text{not-H}) = p(\text{not-D}|\text{not-H}) = 1/2$. This results in

$$\frac{p(\text{H}|\text{D})}{p(\text{not-H}|\text{D})} = 2$$

Because $p(\text{not-H}|\text{D}) = 1 - p(\text{H}|\text{D})$, we get $p(\text{H}|\text{D}) = 2 - 2p(\text{H}|\text{D})$, which is equivalent to

$$3p(\text{H}|\text{D}) = 2;$$

and finally $p(\text{H}|\text{D}) = 2/3$.

2. The rule of succession can explain the false consensus effect (see Dawes, 1990). In one study, students at Stanford University were asked to engage in activities such as walking around the campus wearing a big sign that said "Repent!" After each student had complied or refused, he or she was asked to estimate the proportion of Stanford students who would comply. On average, those who had complied estimated that about 2/3 of the others would also do so; those who refused estimated that only about 1/4 would comply. The difference between the two groups' estimates was considered an egocentric bias because the students guessed what others would do on the basis of their own view of the situation. However, a student most likely never has asked other students whether they would be willing to walk around with this sign, but only knows what he or she responded. If a student's own answer was "comply," then the rule of succession dictates that she or he give a probability of 2/3 that the next student would also comply. This is about equal to the average estimate given by the "compliers" in the study. If the student's

answer was "refuse," then the rule of succession dictates that she or he give a probability of 2/3 that the next student would also refuse, or a probability of 1/3 that the next student would comply. This is in the same direction as the average estimate given by the "refusers" in the study. Thus, the false consensus effect can be seen as the product of a rational process—the same used by Adam and Eve when they predicted that the sun would rise on their second day in Paradise.

3. Hammerton (1973, p. 252).

4. The irony is that Hammerton used what he called a "'commonsense' verbal argument" (p. 252)—which was expressed in terms of natural frequencies—to explain the problem and its solution to the reader, but not the participants in his study.

5. Svenson, Fischhoff, and McGregor (1985, p. 119).

6. The idea that skewed distributions can account for the fact that most drivers report that they are better than average is developed in Lopes (1992) and Schwing and Kamerud (1988).

7. A different way to describe this phenomenon is to say that in the symmetrical distribution of Figure 13-3 (top) the average number (arithmetic mean) of accidents coincides with the median number of accidents. (The median is the point that cuts the distribution into halves, with 50 percent of the drivers above the median and 50 percent below.) In the asymmetrical distribution (bottom) the average and the median are different: the median is 3 accidents, but the arithmetic mean is 4.5. This discrepancy can have puzzling consequences for expected utility maximization theories (which are based on the arithmetic mean rather than the median) when distributions are highly asymmetrical. Examples are the St. Petersburg paradox in humans (Lopes, 1981) and "adaptive coin flipping" in animals (Cooper, 1989; Gigerenzer, 1996b).

8. "Ask Marilyn," *Parade* magazine, September 9, 1990, p. 15 (1990a), and December 2, p. 25 (1990b). The Monty Hall problem was first stated in the *American Statistician* by Steve Selvin (1975a; 1975b).

9. See Granberg and Brown (1995); Krauss and Wang (2000).

10. Tierney (1991). For the three letters quoted in the text, see vos Savant (1996). The second letter was on the three prisoners problem, which is structurally very similar (see next section).

11. Krauss and Wang (2000) derived these methods from the reports of successful participants and showed empirically that perspective change and frequency judgments increased the proportion of people who switch. There are other creative solutions. One person said, "I would walk up to each door and try to smell the goat." The answer to the Monty Hall problem can also be computed using Bayes's rule. Assume the guest picks door 1. The probability $p(\text{car} = 1)$ that the car is behind door 1 is 1/3 and is the same for the other two doors. Now Monty opens door 3 and shows a goat. What is the probability $p(\text{car} = 2 | \text{goat} = 3)$ that the car is behind

door 2 given that Monty opened door 3 and showed a goat? In other words, what is the probability of winning the car if the guest switches from door 1 to door 2? The answer can be calculated as follows: $p(\text{car} = 2|\text{goat} = 3) = p(\text{car} = 2)p(\text{goat} = 3|\text{car} = 2)/[p(\text{car} = 1)p(\text{goat} = 3|\text{car} = 1) + p(\text{car} = 2)p(\text{goat} = 3|\text{car} = 2) + p(\text{car} = 3)p(\text{goat} = 3|\text{car} = 3)] = 1/3 \times 1/[1/3 \times 1/2 + 1/3 \times 1 + 1/3 \times 0] = 2/3$.

12. Friedman (1998).
13. Mueser and Granberg (1999).
14. Early versions of this problem appeared in Gardner (1959a; 1959b) and Mosteller (1965), and Falk (1992) analyzes people's intuitions about this problem.

CHAPTER 14. TEACHING CLEAR THINKING

1. Kohn, Corrigan, and Donaldson (2000).
2. Schönhöfer (1999). This means that roughly one in a thousand hospital patients is killed by medication errors. There are more than 50,000 drugs already on the market, too many for a physician to evaluate, plus a constant stream of new drugs. The number of serious conditions in Germany due to medication errors is estimated at 120,000 to 240,000 per year.
3. Chang (2000, p. 10).
4. Chang (2000, p. 10).
5. For instance, see discussion of *People v. Collins* (Chapter 9).
6. Gigerenzer (1993, 2000).
7. Reimer et al. (1997).
8. In his work of science fiction, *1984,* the writer George Orwell warned that Big Brother would deprive people of their books, autonomy, and freedom. As yet there is no Big Brother to force us to believe in the illusion of certainty or ban us from reasoning about risks. However, another writer of science fiction foresaw a more realistic and disturbing society. In Aldous Huxley's *Brave New World,* no Big Brother is needed to deprive people of information and the ability to think; they simply have no desire to know. Instead of wanting to understand their world, they have an almost boundless appetite for distraction and entertainment. Huxley's work highlights one threat to the teaching of statistical thinking: lack of awareness of uncertainty and the curiosity that comes with it.
9. In his seminal book, *Cancer Wars* (1996), Robert Proctor draws attention to the role of trade associations in generating "doubt." This section draws on his work (pp. 101–110).
10. Proctor (1999).
11. Proctor (1998).
12. Proctor (1998, p. 10).
13. Proctor (1996, p. 107).

14. Proctor (1996, p. 108).
15. Proctor (1996, p. 110). The production of confusion and ignorance concerning the health hazards of asbestos is strikingly similar (Proctor, 1996, pp. 110–122).
16. Dehaene (1997, p. 115).
17. E.g., Douglas and Wildavsky (1982); Fischhoff et al. (1981); Lopes (1987; 1992); Slovic (1987); Tooby and Cosmides (1992).
18. Mineka and Cook (1988).
19. Hazards that have catastrophic potential, can get out of control, and have fatal consequences are often called "dread" risks (Slovic, 1987).
20. Reiss and Roth (1993).
21. National Research Council (1996).
22. U.S. Preventive Services Task Force Staff (1996).
23. Center for the Evaluative Clinical Sciences Staff (1996).
24. Krauss, Martignon, and Hoffrage (1999).
25. Sedlmeier (1999); Sedlmeier and Gigerenzer (2001)
26. For details see Sedlmeier and Gigerenzer (2001).
27. Daston (1988).
28. See Gigerenzer (2000, Chapter 12).

REFERENCES

Adler, I. (1912). *Primary malignant growths of lungs and bronchi.* New York: Longmans, Green, and Co.

Altman, L. K. (2000, July 16). Africa's AIDS crisis: Finding common ground. *The New York Times,* p. 4.

Antell, S. E., & Keating, D. P. (1983). Perception of numerical invariants in neonates. *Child Development, 54,* 695–701.

Atkinson, D. (1977). Society and sexes in the Russian past. In D. Atkinson, A. Dallin, & G. W. Lapidus (Eds.), *Women in Russia* (pp. 3–38). Stanford, CA: Stanford University Press.

Baines, C. J. (1992). Women and breast cancer: Is it really possible for the public to be well informed? *The Canadian Medical Association Journal, 142,* 2147–2148.

Balding, D. J., & Donnelly, P. (1994). How convincing is DNA evidence? *Nature, 368,* 285–286.

Barsalou, L. W., & Ross, B. H. (1986). The roles of automatic and strategic processing in sensitivity to superordinate and property frequency. *Journal of Experimental Psychology: Learning, Memory, and Cognition, 12,* 116–134.

Beck, U. (1986) *Risikogesellschaft: Auf dem Weg in eine andere Moderne.* Frankfurt am Main: Suhrkamp.

Bernard, C. (1865/1957). *An introduction to the study of experimental medicine* (H. C. Greene, Trans.). New York: Dover.

Berner, E. S. (1997). When to teach Bayesian reasoning (Letter to the editor). *Journal of Medical Decision Making, 17,* 233.

Berry, D. A. (1991). Inferences using DNA profiling in forensic identification and paternity cases. *Statistical Science, 6,* 175–205.

Berwick, D. M., Fineberg, H. V., & Weinstein, M. C. (1981). When doctors meet numbers. *American Journal of Medicine, 71,* 991–998.

Black, W. C., Nease, R. F., Jr., & Tosteson, A. N. A. (1995). Perceptions of breast cancer risk and screening effectiveness in women younger than 50 years of age. *Journal of the National Cancer Institute, 87,* 720–731.

Bondy, K. (1999). *Hack Barbie!* [On-line]. Available: http://www.zdnet.com/zdtv/thesite/0397w4/view/iview431jump2_031797.html.

Bourguet, M.-N. (1987). Décrire, compter, calculer: The debate over statistics during the Napoleonic period. In L. Krüger, L. Daston, & M. Heidelberger (Eds.), *The probabilistic revolution: Vol. I. Ideas in history* (pp. 305–316). Cambridge, MA: MIT Press.

Boyer, P. J. (2000, January 17). Annals of justice: DNA on trial. *The New Yorker*, 42–53.

Brase, G. L., Cosmides, L., & Tooby, J. (1998). Individuation, counting, and statistical inference: The role of frequency and whole object representations in judgment under uncertainty. *Journal of Experimental Psychology: General, 127*, 3–21.

Breast cancer [Cover story]. (1991). *Time, 137*, 42–49.

Brunswik, E. (1937). Psychology as a science of objective relations. *Philosophy of Science, 4*, 227–260.

Bundesamt für Gesundheit (BAG) in Zusammenarbeit mit der Eidgenössischen Kommission für Aids-Fragen (EKAF). (2000). *Informationen zum HIV-Test*. Bern: Bundesamt für Gesundheit, Schweiz.

Bundeszentrale für gesundheitliche Aufklärung (Ed.). (1988–1993). *Wissenswertes über den HIV-Test, Issues 1–10*. Köln: Bundeszentrale für gesundheitliche Aufklärung.

Bundeszentrale für gesundheitliche Aufklärung (Ed.). (1993). *Handbuch HIV-Test. Arbeitshilfen zur Beratung und Testdurchführung*. Köln: Bundeszentrale für gesundheitliche Aufklärung.

Bundeszentrale für gesundheitliche Aufklärung (Ed.). (2000). *Leben mit HIV: Wenn der HIV-Test positiv ist. Informationen und Orientierungshilfe*. Köln: Bundeszentrale für gesundheitliche Aufklärung.

Bursztajn, H., Feinbloom, R. I., Hamm, R. M., & Brodsky, A. (1981). *Medical choices, medical chances: How patients, families, and physicians can cope with uncertainty*. New York: Delta/Seymour Lawrence.

Busch, M. P. (1994). HIV testing in blood banks. In G. Schochetman & J. R. George (Eds.), *AIDS testing: A comprehensive guide to technical, medical, social, legal, and management issues* (pp. 224–236). New York: Springer.

Butterworth, B. (1999). *The mathematical brain*. London: Macmillan.

Casscells, W., Schoenberger, A., & Grayboys, T. (1978). Interpretation by physicians of clinical laboratory results. *New England Journal of Medicine, 299*, 999–1000.

Catalan, J., & Pugh, K. (1995). Suicidal behaviour and HIV infection—is there a link? *AIDS Care, 7*, S117–S121.

Center for the Evaluative Clinical Sciences Staff (Ed.). (1996). *The Dartmouth atlas of health care*. Chicago: American Hospital Association.

Chang, K. (2000, May 4). Findings fuel debate over prostate testing. *International Herald Tribune*, p. 10.

Chivers, C. J. (2000, February 9). As DNA aids rape inquiries, statutory limits block cases. *The New York Times*, late edition, section B, p. 1.

Cockburn, J., Redman, S., Hill, D., & Henry, E. (1995). Public understanding of medical screening. *Journal of Medical Screening, 2*, 224–227.

Coleman, W. (1987). Experimental physiology and statistical inference: The therapeutic trial in nineteenth-century Germany. In L. Krüger, G. Gigerenzer, & M. S. Morgan (Eds.), *The probabilistic revolution: Vol. II. Ideas in the sciences* (pp. 201–226). Cambridge, MA: MIT Press.

Collins, R., & Macleod, A. (1991). Denials of paternity: The impact of DNA tests on court proceedings. *The Journal of Social Welfare and Family Law, 3,* 209–219.

Cooper, W. S. (1989). How evolutionary biology challenges the classical theory of rational choice. *Biology and Philosophy, 4,* 457–481.

Cosmides, L., & Tooby, J. (1996). Are humans good intuitive statisticians after all? Rethinking some conclusions from the literature on judgment under uncertainty. *Cognition, 58,* 1–73.

Cresanta, J. L. (1992). Epidemiology of cancer in the United States. *Cancer Epidemiology, Prevention, and Screening, 19,* 419–441.

Daly, M., & Wilson, M. (1988). *Homicide.* New York: Aldine de Gruyter.

Daston, L. (1981). Mathematics and the moral sciences: The rise and fall of the probability of judgments, 1785–1840. In H. N. Jahnke & M. Otte (Eds.), *Epistemological and social problems of the sciences in the early nineteenth century* (pp. 287–309). Dordrecht, Holland: D. Reidel Publishing Company.

Daston, L. (1987). The domestication of risk: Mathematical probability and insurance 1650–1830. In L. Krüger, L. Daston, & M. Heidelberger (Eds.), *The probabilistic revolution, Vol. 1: Ideas in history* (pp. 237–260). Cambridge, MA: MIT Press.

Daston, L. (1988). *Classical probability in the Enlightenment.* Princeton, NJ: Princeton University Press.

Dawes, R. M. (1986). Representative thinking in clinical judgment. *Clinical Psychology Review, 6,* 425–441.

Dawes, R. M. (1990). The potential nonfalsity of the false consensus effect. In R. M. Hogarth (Ed.), *Insights in decision making: A tribute to Hillel J. Einhorn* (pp. 179–199). Chicago: The University of Chicago Press.

Dawes, R. M. (1994). *House of cards: Psychology and psychotherapy built on myth.* New York: The Free Press.

Dawes, R. M. (2001). *Everyday irrationality.* Boulder, CO: Westview.

Dehaene, S. (1997). *The number sense: How the mind creates mathematics.* New York: Oxford University Press.

A delicate question. (1999, 24 April). *The Economist, 26.*

Dershowitz, A. M. (1983). *The best defense.* New York: Vintage Books.

Dershowitz, A. M. (1996). *Reasonable doubts: The criminal justice system and the O. J. Simpson case.* New York: Simon and Schuster.

Deutscher Bundestag (Ed.). (1990). *AIDS: Fakten und Konsequenzen. Final report of the Enquete Committee of the 11th German Bundestag, 13/90.* Bonn: Bonner Universitäts Buchdruckerei.

Dolan, N. C., Lee, A. M., & McDermott, M. M. (1997). Age-related differences in breast carcinoma knowledge, beliefs, and perceived risk among women visiting an academic general medicine practice. *Cancer, 80,* 413–420.

Domenighetti, J. (2000). *General public perception of mammography screening benefits.* Paper presented at the Einsiedler Symposium, Einsiedeln, Switzerland.

Douglas, M. (1992). *Risk and blame: Essays in cultural theory.* London: Routledge.

Douglas, M., & Wildavsky, A. (1982). *Risk and culture: An essay on the selection of technological and environmental dangers.* Berkeley, CA: University of California Press.

Earman, J. (1992). *Bayes or bust? A critical examination of Bayesian confirmation theory.* Cambridge, MA: MIT Press.

Eddy, D. M. (1982). Probabilistic reasoning in clinical medicine: Problems and opportunities. In D. Kahneman, P. Slovic, & A. Tversky (Eds.), *Judgment under uncertainty: Heuristics and biases* (pp. 249–267). Cambridge: Cambridge University Press.

Eddy, D. M. (1996). *Clinical decision making: From theory to practice: A collection of essays from the Journal of the American Medical Association.* Boston: Jones and Bartlett Publishers.

Ellis, G. K. (1994). Oncology update: Breast cancer. *Primary Care Update for OB/GYNS, 1,* 17–25.

Elmore, J. G., Barton, M. B., Moceri, V. M., Polk, S., Arena, P. J., & Fletcher, S. W. (1998). Ten-year risk of false positive screening mammograms and clinical breast examinations. *The New England Journal of Medicine, 338,* 1089–1096.

Ernster, V. L., & Barclay, J. (1997). Increases in ductal carcinoma *in situ* (DCIS) of the breast in relation to mammography: A dilemma. *Journal of the National Cancer Institute/Monographs, 22,* 151–156.

Estes, W. K. (1976). The cognitive side of probability learning. *Psychological Review, 83,* 37–64.

Evers, M., Dworschak, M., Hackenbroch, V., Jaeger, U., Leick, R., Neubacher, A., Schmid, B., & Schreiber, S. (2000, 20 November). Deutschland—Ein BSE-Risikostaat: Seuche aus dem Trog. *Der Spiegel, 47,* 288–292.

Evers, M., Neubacher, A., Pötzl, N., Schreiber, S., & Vehlewald, H.-J. (2000, 27 November). Der deutsche Wahn: Jahrelang haben Politiker und Bauern die Bürger in dem Glauben gelassen, ihr Land sei frei von der Rinderseuche BSE. *Der Spiegel, 48,* 22–24.

Fahey, T., Griffiths, S., & Peters, T. J. (1995). Evidence based purchasing: Understanding results of clinical trials and systematic reviews. *British Medical Journal, 311,* 1056–1059.

Faigman, D. L. (1999). *Legal alchemy: The use and misuse of science in the law.* New York: W. H. Freeman and Company.

Falk, R. (1992). A closer look at the probabilities of the notorious three prisoners. *Cognition, 43,* 197–223.

Fawcett, G. M., Heise, L. L., Isita-Espejel, L., & Pick, S. (1999). Changing community responses to wife abuse: A research and demonstration project in Iztacalco, Mexico. *American Psychologist, 54,* 41–49.

Feynman, R. P. (1967). *The character of physical law.* Cambridge, MA: MIT Press.

Fingering fingerprints. (2000, December 16). *The Economist, 357,* 103–104.

Fischhoff, B. (1982). For those condemned to study the past: Heuristics and biases in hindsight. In D. Kahneman, P. Slovic, & A. Tversky (Eds.), *Judgment under uncertainty: Heuristics and biases* (pp. 335–351). Cambridge: Cambridge University Press.

Fischhoff, B., Lichtenstein, S., Slovic, P., Darby, S., & Keeney, R. (1981). *Acceptable risk.* New York: Cambridge University Press.

Fisher, R. A. (1935). *The design of experiments.* (5th ed., 1951; 7th ed., 1960; 8th ed., 1966). Edinburgh: Oliver & Boyd.

Franklin, B. (1987). *Writings.* New York: The Library of America.

Friedman, D. (1998). Monty Hall's three doors: Construction and deconstruction of a choice anomaly. *The American Economic Review, 88,* 933–946.

Gallistel, C. R., & Gelman, R. (1992). Preverbal and verbal counting and computation. *Cognition, 44,* 43–74.

Gardner, M. (1959a). Mathematical games: How three mathematicians disproved a celebrated conjecture of Leonard Euler. *Scientific American, 201,* 181–188.

Gardner, M. (1959b). Mathematical games: Problems involving questions of probability and ambiguity. *Scientific American, 201,* 147–182.

Garnick, M. B. (1994). The dilemmas of prostate cancer. *Scientific American, 270,* 52–59.

Gawande, A. (1999, October 4). Whose body is it, anyway? What doctors should do when patients make bad decisions. *The New Yorker,* 84–91.

Geary, D. C. (2000). Evolution and proximate expression of human parental investment. *Psychological Bulletin, 126,* 55–77.

Gigerenzer, G. (1993). The superego, the ego, and the id in statistical reasoning. In G. Keren & G. Lewis (Eds.), *A handbook for data analysis in the behavioral sciences* (pp. 311–339). Hilsdale, NJ: Erlbaum.

Gigerenzer, G. (1996a). The psychology of good judgment: Frequency formats and simple algorithms. *Medical Decision Making, 16,* 273–280.

Gigerenzer, G. (1996b). Rationality: Why social context matters. In P. B. Baltes & U. M. Staudinger (Eds.), *Interactive minds: Life-span perspectives on the social foundation of cognition* (pp. 319–346). Cambridge: Cambridge University Press.

Gigerenzer, G. (1998). Ecological intelligence: An adaptation for frequencies. In D. D.

Cummins & C. Allen (Eds.), *Evolution of Mind* (pp. 9–29). New York: Oxford University Press.

Gigerenzer, G. (2000). *Adaptive thinking: Rationality in the real world.* New York: Oxford University Press.

Gigerenzer, G., & Hoffrage, U. (1995). How to improve Bayesian reasoning without instruction: Frequency formats. *Psychological Review, 102,* 684–704.

Gigerenzer, G., & Hoffrage, U. (1999). Overcoming difficulties in Bayesian reasoning: A reply to Lewis and Keren (1999) and Mellers and McGraw (1999). *Psychological Review, 106,* 425–430.

Gigerenzer, G., Hoffrage, U., & Ebert, A. (1998). AIDs counselling for low-risk clients. *AIDS Care, 10,* 197–211.

Gigerenzer, G., Hoffrage, U., & Kleinbölting, H. (1991). Probabilistic mental models: A Brunswikian theory of confidence. *Psychological Review, 98,* 506–528.

Gigerenzer, G., & Murray, D. J. (1987). *Cognition as intuitive statistics.* Hillsdale, NJ: Erlbaum.

Gigerenzer, G., Swijtink, Z., Porter, T., Daston, L., Beatty, J., & Krüger, L. (1989). *The empire of chance. How probability changed science and everyday life.* Cambridge: Cambridge University Press.

Gigerenzer, G., Todd, P. M. & the ABC Research Group (1999). *Simple heuristics that make us smart.* New York: Oxford University Press.

Gillispie, C. C. (1997). *Pierre-Simon Laplace, 1749–1827: A life in exact science.* Princeton, NJ: Princeton University Press.

Good, I. J. (1995). When batterer turns murderer. *Nature, 375,* 541.

Good, I. J. (1996). When batterer becomes murderer. *Nature, 381,* 481.

Gøtzsche, P. C., & Olsen, O. (2000). Is screening for breast cancer with mammography justifiable? *The Lancet, 355,* 129–134.

Granberg, D., & Brown, T. A. (1995). The Monty Hall dilemma. *Personality and Social Psychology Bulletin, 21,* 711–723.

Grisso, T., & Tomkins, A. J. (1996). Communicating violence risk assessments. *American Psychologist, 51,* 928–930.

Größter Massen-Gentest ist noch nicht beendet: Der Mordfall Christina / Streit unter Rechtsmedizinern. (1998, April 14). *Frankfurter Allgemeine Zeitung,* p. 13.

Hacking, I. (1975). *The emergence of probability.* Cambridge: Cambridge University Press.

Hacking, I. (1990). *The taming of chance.* Cambridge: Cambridge University Press.

Haley, N. J., & Reed, B. S. (1994). HIV testing for life insurance. In G. Schochetman & J. R. George (Eds.), *AIDS testing: A comprehensive guide to technical, medical, social, legal, and management issues* (2nd ed., pp. 252–265). New York: Springer.

Hamm, R. M., Lawler, F., & Scheid, D. (1999). Prophylactic mastectomy in women with a high risk of breast cancer. *New England Journal of Medicine, 340,* 1837–1838.

Hamm, R. M., & Smith, S. L. (1998). The accuracy of patients' judgments of disease probability and test sensitivity and specificity. *The Journal of Family Practice, 47,* 44–52.

Hammerton, M. (1973). A case of radical probability estimation. *Journal of Experimental Psychology, 101,* 252–254.

Hanks, G. E., & Scardino, P. T. (1996). Does screening for prostate cancer make sense? *Scientific American, 275,* 80–81.

Hansen, M. (1996, August). Jimmy the Greek he ain't. *American Bar Association Journal, 30.*

Harrington, A. (1997). Placebo: Conversations at the disciplinary borders. In A. Harrington (Ed.), *The placebo effect: An interdisciplinary exploration* (pp. 208–248). Cambridge, MA: Harvard University Press.

Harris, R. P., Fletcher, S. W., Gonzalez, J. J., Lannin, D. R., Degnan, D., Earp, J. A., & Clark, R. (1991). Mammography and age: Are we targeting the wrong women? *Cancer, 67,* 2010–2014.

Hartmann, L. C. et al. (1999). Efficacy of bilateral prophylactic mastectomy in women with a family history of breast cancer. *The New England Journal of Medicine, 340,* 77–84.

Hasher, L., & Zacks, R. T. (1984). Automatic processing of fundamental information: The case of frequency of occurrence. *American Psychologist, 39,* 1372–1388.

Heilbrun, K., O'Neill, M. L., Strohman, L. K., Bowman, Q., & Philipson, J. (2000). Expert approaches to communicating violence risk. *Law and Human Behavior, 24,* 137–148.

Hicks, J. W. (1993). The facts about DNA typing. *Judicature, 77,* 5, 55, 57–58.

Hippocrates. (1967). Decorum, *Hippocrates* (Vol. II, pp. 297–299). Cambridge, MA: Harvard University Press.

Hoffrage, U., & Gigerenzer, G. (1998). Using natural frequencies to improve diagnostic inferences. *Academic Medicine, 73,* 538–540.

Hoffrage, U., Lindsey, S., Hertwig, R., & Gigerenzer, G. (2000). Communicating statistical information. *Science, 290,* 2261–2262.

Horne, S. (1999). Domestic violence in Russia. *American Psychologist, 54,* 55–61.

Huff, D. (1954/1993). *How to lie with statistics.* New York: W. W. Norton & Company.

Humphrey, N. (2000). *Great expectations: The evolutionary psychology of faith-healing and the placebo effect* [Keynote address]. Paper presented at the XXVII International Congress of Psychology, Stockholm.

Illinois Department of Public Health (1992). *Coping with HIV disease.* San Francisco AIDS Foundation.

Illinois Department of Public Health (1993). *AIDS: Antibody testing.* (Leaflet). Printed by Authority of the State of Illinois. P.O. X302237 90,200 1/93.

Jain, B. P., McQuay, H., & Moore, A. (1998). Number needed to treat and relative risk reduction. *Annals of Internal Medicine, 128,* 72–73.

Jasanoff, S., & Lynch, M. (1998). Contested identities: Science, law, and forensic practice. *Social Studies of Science, 28,* 675–686.

Jonides, J., & Jones, C. M. (1992). Direct coding for frequency of occurrence. *Journal of Experimental Psychology: Learning, Memory, and Cognition, 18,* 368–378.

Jung, H. (1998). Mammographie und Strahlenrisiko. *Fortschritte auf dem Gebiet der Röntgenstrahlen, 169,* 336–343.

Kahneman, D., Slovic, P., & Tversky, A. (1982). *Judgment under uncertainty: Heuristics and biases.* New York: Cambridge University Press.

Kalet, A., Roberts, J. C., & Fletcher, R. (1994). How do physicians talk with their patients about risks? *Journal of General Internal Medicine, 9,* 402–404.

Kant, I. (1784). Beantwortung der Frage: Was ist Aufklärung? *Berlinische Monatsschrift, Dezember-Heft,* 481–494.

Katz, J. (1984). *The silent world of doctor and patient.* New York: The Free Press.

Kerlikowske, K. (1997). Efficacy of screening mammography among women aged 40 to 49 years and 50 to 69 years: Comparison of relative and absolute benefit. *Journal of the National Cancer Institute Monographs, 22,* 79–86.

Kerlikowske, K. (2000). Breast cancer screening. In M. B. Goldman & M. C. Hatch (Eds.), *Women and health* (pp. 895–906). New York: Academic Press.

Kerlikowske, K., Grady, D., Barclay, J., Sickles, E. A., & Ernster, V. (1996a). Effect of age, breast density, and family history on the sensitivity of first screening mammography. *Journal of the American Medical Association, 276,* 33–38.

Kerlikowske, K., Grady, D., Barclay, J., Sickles, E. A., & Ernster, V. (1996b). Likelihood ratios for modem screening mammography: Risk of breast cancer based on age and mammographic interpretation. *Journal of the American Medical Association, 276,* 39–43.

Kleiter, G. D. (1994). Natural sampling: Rationality without base rates. In G. H. Fischer & D. Laming (Eds.), *Contributions to mathematical psychology, psychometrics, and methodology* (pp. 375–388). New York: Springer.

Knight, F. (1921). *Risk, uncertainty and profit.* Boston: Houghton Mifflin Co.

Koehler, J. J. (1992). Probabilities in the courtroom: An evaluation of the objections and policies. In D. K. Kagehiro, & W. S. Laufer (Eds.), *Handbook of psychology and law* (pp. 167–184). New York: Springer.

Koehler, J. J. (1993a). DNA matches and statistics: Important questions, surprising answers. *Judicature, 76,* 222–229.

Koehler, J. J. (1993b). Error and exaggeration in the presentation of DNA evidence at trial. *Jurimetrics Journal, 34,* 21–39.

Koehler, J. J. (1996). On conveying the probative value of DNA evidence: Frequencies, likelihood ratios, and error rates. *University of Colorado Law Review, 67,* 859–886.

Koehler, J. J. (1997). One in millions, billions, and trillions: Lessons from *People vs.*

Collins (1968) for *People vs. Simpson* (1995). *Journal of Legal Education, 47,* 214–223.

Koehler, J. J., Chia, A., & Lindsey, S. (1995). The random match probability (RMP) in DNA evidence: Irrelevant and prejudicial? *Jurimetrics Journal, 35,* 201–219.

Kohn, L. T., Corrigan, J. M., & Donaldson, M. S. (Eds.). (2000). *To err is human: Building a safer health system.* Washington, DC: National Academy Press.

Koss, M. P., Koss, P. G., & Woodruff, J. (1991). Deleterious effects of criminal victimization on women's health and medical utilization. *Archives of Internal Medicine, 151,* 342–347.

Koubenec, H.-J. (2000). Mammographie-Screening: Überschätzen wir den Nutzen? *Berliner Ärzte, 8,* 11–16.

Krauss, S., & Hertwig, R. (2000). Muss DNA-Evidenz schwer verständlich sein? Der Ausweg aus einem Kommunikationsproblem. *Monatsschrift für Kriminologie und Strafrechtsreform, 83,* 155–162.

Krauss, S., Martignon, L., & Hoffrage, U. (1999). Simplifying Bayesian inference: The general case. In L. Magnani, N. Nersessian, & N. Thagard (Eds.), *Model-based reasoning in scientific discovery* (pp. 165–179). New York: Plenum Press.

Krauss, S., & Wang, X. T. (2000). *The psychology of the Monty Hall problem: Overcoming difficulties in solving a tenacious brain teaser.* Unpublished manuscript.

Krüger, L., Daston, L., & Heidelberger, M. (1987). *The probabilistic revolution: Vol. 1. Ideas in history.* Cambridge, MA: MIT Press.

Krüger, L., Gigerenzer, G., & Morgan, M. S. (Eds.). (1987). *The probabilistic revolution: Vol. II. Ideas in the sciences.* Cambridge, MA: MIT Press.

Kühberger, A. (1998). The influence of framing on risky decisions: A meta-analysis. *Organizational Behavior and Human Decision Processes, 75,* 23–55.

Lantz, P. M., & Booth, K. M. (1998). The social construction of the breast cancer epidemic. *Social Science and Medicine, 46,* 907–918.

Laplace, P.-S. (1814/1951). *A philosophical essay on probabilities* (F. W. Truscott and F. L. Emory, Trans.). New York: Dover.

Lempert, R. (1991). Some caveats concerning DNA as criminal identification evidence: With thanks to the Reverend Bayes. *Cardozo Law Review, 13,* 303–341.

Lerman, C., Trock, B., Rimer, B. K., Jepson, C., Brody, D., & Boyce, A. (1991). Psychological side effects of breast cancer screening. *Health Psychology, 10,* 259–267.

LeRoy, S. F., & Singell, L. D., Jr. (1987). Knight on risk and uncertainty. *Journal of Political Economy, 95,* 394–406.

Lewis, C., & Keren, G. (1999). On the difficulties underlying Bayesian reasoning: A comment on Gigerenzer and Hoffrage. *Psychological Review, 106,* 411–416.

Lindsey, S., Hertwig, R., & Gigerenzer, G. (2001). *Communicating DNA evidence.* Unpublished manuscript.

Lopes, L. L. (1981). Decision making in the short run. *Journal of Experimental Psychology: Human Learning and Memory, 7,* 377–385.

Lopes, L. L. (1987). Between hope and fear: The psychology of risk. In L. Berkowitz (Ed.), *Advances in Experimental Social Psychology* (Vol. 20, pp. 255–295). San Diego: Academic Press.

Lopes, L. L. (1992). Risk perception and the perceived public. In D. W. Bromley & K. Segerson (Eds.), *The social response to environmental risk* (pp. 57–73). Boston: Kluwer Academic Publishers.

Malenka, D. J., Baron, J. A., Johansen, S., Wahrenberger, J. W., & Ross, J. M. (1993). The framing effect of relative and absolute risk. *Journal of General Internal Medicine, 8,* 543–548.

Mandel, J. S., et al. (1993). Reducing mortality from colorectal cancer by screening for fecal occult blood. *New England Journal of Medicine, 328,* 1365–1371.

Månsson, S. A. (1990). Psycho-social aspects of HIV testing—the Swedish case. *AIDS Care, 2,* 5–16.

Marr, D. (1982). *Vision: A computational investigation into the human representation and processing of visual information.* San Francisco: W.H. Freeman.

Marshall, E. (1993). Search for a killer: Focus shifts from fat to hormones. *Science, 259,* 618–621.

Marshall, K. G. (1996). Prevention. How much harm? How much benefit? 3. Physical, psychological and social harm. *Canadian Medical Association Journal, 155,* 169–176.

Matthews, J. R. (1995). *Quantification and the quest for medical certainty.* Princeton, NJ: Princeton University Press.

Mazur, D. J., & Metz, J. F. (1994). How age, outcome severity, and scale influence general medicine clinic patients' interpretations of verbal probability terms. *Journal of General Internal Medicine, 9,* 268–271.

McKenzie, C. R. (1994). The accuracy of intuitive judgment strategies: Covariation assessment and Bayesian inference. *Cognitive Psychology, 26,* 209–239.

McNeil, B. J., Pauker, S. G., Sox, H. C., & Tversky, A. (1982). On the elicitation of preferences for alternative theories. *New England Journal of Medicine, 306,* 1259–1262.

McWhirter, P. T. (1999). La violencia privada: Domestic violence in Chile. *American Psychologist, 54,* 37–40.

Metsch, L. R., McCoy, C. B., McCoy, H. V., Pereyra, M., Trapido, E., & Miles, C. (1998). The role of the physician as an information source on mammography. *Cancer Practice, 6,* 229–236.

Miller, K., Smith, C., Zhu, J., & Zhang, H. (1995). Preschool origins of cross-national differences in mathematical competence: The role of number-naming systems. *Psychological Science, 6,* 56–60.

Mineka, S., & Cook, M. (1988). Social learning and the acquisition of snake fear in monkeys. In T. R. Zentall & B. G. Galef (Eds.), *Social learning: Psychological and biological perspectives* (pp. 51–73). Hillsdale, NJ: Erlbaum.

Monahan, J. (1981). *The clinical prediction of violent behavior.* Washington, DC: U.S. Government Printing House.

Monahan, J., Steadman, H. J., Silver, E., Appelbaum, P. S., Robbins, P. C., Mulvey, E. P., Roth, L. H., Grisso, T., & Banks, S. (2001). *Rethinking risk assessment: The McArthur study of mental disorder and violence.* New York: Oxford University Press.

Monahan, J., & Wexler, D. B. (1978). A definite maybe: Proof and probability in civil commitment. *Law and Human Behavior, 2,* 37–42.

Mosteller, F. (1965). *Fifty challenging problems in probability with solutions.* Reading, MA: Addison-Wesley.

Mueser, P. R., & Granberg, D. (1999). The Monty Hall dilemma revisited: Understanding the interaction of problem definition and decision making. Under revision.

Mühlhauser, I., & Höldke, B. (1999). Übersicht: Mammographie-Screening—Darstellung der wissenschaftlichen Evidenz-Grundlage zur Kommunikation mit der Frau. *Sonderbeilage arznei-telegramm, 10/99,* 101–108.

Mushlin, A. I., Kouides, R. W., & Shapiro, D. E. (1998). Estimating the accuracy of screening mammography: A meta-analysis. *American Journal of Preventive Medicine, 14,* 143–153.

National Academy of Sciences Committee on the biological effects of ionizing radiations. (1990). *Health effects of exposure to low levels of radiations.* (BEIR V ed.). Washington, DC: National Academy Press.

The National Commission on Excellence in Education. (1983). *A nation at risk: The imperative for educational reform* (A Report to the Nation and the Secretary of Education, United States Department of Education). Washington, DC: U.S. Government Printing Office.

National Research Council. (1989). *Improving risk communication.* Washington, DC: National Academy Press.

National Research Council. (1996). *The evaluation of forensic DNA evidence. Committee on DNA forensic science: An update.* Washington, DC: National Academy of Sciences.

Nesse, R. M., & Williams, G. C. (1995). *Why we get sick: The new science of Darwinian medicine.* New York: Vintage Books.

NIH Consensus Statement. (1997). Breast cancer screening for women ages 40–49. *NIH Consensus Statement, 15,* 1–35.

Nyström, L., Larsson, L.-G., Wall, S., Rutqvist, L., Andersson, I., Bjurstam, N., Fagerberg, G., Frisell, J., & Tabár, L. (1996). An overview of the Swedish randomised mammography trials: Total mortality pattern and the representativity of the study cohorts. *Journal of Medical Screening, 3,* 85–87.

Oberlies, D. (1997). Tötungsdelikte zwischen Männern und Frauen: Eine Untersuchung geschlechtsspezifischer Unterschiede anhand von 174 Gerichtsurteilen. *Monatsschrift für Kriminologie und Strafrechtsreform, 80,* 133–147

Olsen, O. & Gøtzsche, P. C. (2001). Cochrane review on screening for breast cancer with mammography. *Lancet, 358,* 1340–1342.

Parducci, A. (1965). Category judgment: A range-frequency model. *Psychological Review, 72,* 407–418.

Paulos, J. A. (1988). *Innumeracy: Mathematical illiteracy and its consequences.* New York: Vintage Books.

The perils of percentages. (1998, April 18). *The Economist,* 84.

Phillips, K.-A., Glendon, G., & Knight, J. A. (1999). Putting the risk of breast cancer in perspective. *New England Journal of Medicine, 340,* 141–144.

Politser, P. E. (1984). Explanations of statistical concepts: Can they penetrate the haze of Bayes? *Methods of Information in Medicine, 23,* 99–108.

Pomata, G. (1998). *Contracting a cure: Patients, healers, and the law in early modern Bologna* (Translated by the author, with the assistance of Rosemarie Foy and Anna Taraboletti-Segre, Trans.). Baltimore: The Johns Hopkins University Press.

Porter, T. (1986). *The rise of statistical thinking, 1820–1900.* Princeton, NJ: Princeton University Press.

Proctor, R. N. (1996). *Cancer wars: How politics shapes what we know and don't know about cancer.* New York: Basic Books.

Proctor, R. N. (1998). Tobacco testimony. Manuscript.

Proctor, R. N. (1999). *The Nazi war on cancer.* Princeton, NJ: Princeton University Press.

Ptacek, J. (1999). *Battered women in the courtroom: The power of judicial responses.* Boston: Northeastern University Press.

Ransohoff, D. F., & Harris, R. P. (1997). Lessons from the mammography screening controversy: Can we improve the debate? *Annals of Internal Medicine, 127,* 1029–1034.

Reagan, N., & Novak, W. (1989). *My turn: The memoirs of Nancy Reagan.* New York: Random House.

Real, L. A. (1991). Animal choice behavior and the evolution of cognitive architecture. *Science, 253,* 980–986.

Redmayne, M. (1998). The DNA database: Civil liberty and evidentiary issues. *Criminal Law Review,* 437–454.

Redmayne, M. (2001). *Expert evidence and criminal justice.* New York: Oxford University Press.

Reimer, L., Mottice, S., Schable, C., Sullivan, P., Nakashima, A., Rayfield, M., Den, R., & Brokopp, C. (1997). Absence of detectable antibody in a patient infected with human immunodeficiency virus. *Clinical Infectious Diseases, 25,* 98–100.

Reiss, A. J., Jr., & Roth, J. A. (Eds.). (1993). *Understanding and preventing violence.* Washington, DC: National Academy Press.

Roberts, M. M. (1989). Breast screening: Time for a rethink? *British Medical Journal, 299,* 1153–1155.

Ross, J. F. (1999). *The polar bear strategy: Reflections on risk in modern life.* Reading, MA: Perseus Books.

Salzmann, P., Kerlikowske, K., & Phillips, K. (1997). Cost-effectiveness of extending screening mammography guidelines to include women 40 to 49 years of age. *Annals of Internal Medicine, 127,* 955–965.

Schindele, E., & Stollorz, V. (2000, March 17). Vorsicht Vorsorge! *Die Woche.*

Schmidt, J. G. (1990). The epidemiology of mass breast cancer screening—A plea for a valid measure of benefit. *Journal of Clinical Epidemiology, 43,* 215–225.

Schmidt, J. G. (1994). Wie gross ist der Nutzen, wie gross der Schaden der Brustkrebs-Früherkennung? Früherkennungs-Credo gegenüber Wirklichkeit in der Praxis. In J. G. Schmidt & R. E. Steele (Eds.), *Kritik der medizinischen Vernunft: Schritte zu einer zeitgemäßen Praxis—Ein Lesebuch* (pp. 63–71). Mainz: Verlag Kirchheim.

Schochetman, G., & George, J. R. (Eds.). (1994). *AIDS testing: A comprehensive guide to technical, medical, social, legal, and management issues* (2nd ed.). New York: Springer.

Schönhöfer, P. S. (1999). Klinik-basierte Erfassung Arzneimittel-bedingter Erkrankungen in Pharmakovigilanz-System (ZKH Bremen). *Arzneimitteltherapie, 17,* 83–86.

Schrage, G. (1980). Schwierigkeiten mit stochastischer Modellbildung—Zwei Beispiele aus der Praxis. *Journal für Mathematik-Didaktik, 1,* 86–88.

Schwartz, L. M., Woloshin, S., Sox, H. C., Fischhoff, B., & Welch, H. G. (2000). U.S. women's reactions to false positive mammography results and detection of ductal carcinoma in situ: Cross-sectional survey. *British Medical Journal, 320,* 1635–1640.

Schwarz, N. (1999). Self-reports: How the questions shape the answers. *American Psychologist, 54,* 93–105.

Schwarz, N., & Hippler, H.-J. (1987). What response scales may tell your respondents: Informative functions of response alternatives. In H.-J. Hippler, N. Schwarz, & S. Sudman (Eds.), *Social information processing survey methodology* (pp. 163–178). New York: Springer.

Schwarz, N., Hippler, H.-J., Deutsch, B., & Strack, F. (1985). Response categories: Effects on behavioral reports and comparative judgments. *Public Opinion Quarterly, 49,* 388–395.

Schwing, R. C., & Kamerud, D. B. (1988). The distribution of risks: Vehicle occupant fatalities and time of week. *Risk Analysis, 8,* 127–133.

Sedlmeier, P. (1999). *Improving statistical reasoning: Theoretical models and practical implications.* Mahwah, NJ: Erlbaum.

Sedlmeier, P., & Gigerenzer, G. (2001). Teaching Bayesian reasoning in less than two hours. *Journal of Experimental Psychology: General, 130,* 380–400.

Sedlmeier, P., Hertwig, R., & Gigerenzer, G. (1998). Are judgments of the positional frequencies of letters systematically biased due to availability? *Journal of Experimental Psychology: Learning, Memory, and Cognition, 24,* 754–770.

Selvin, S. (1975a). A problem in probability. *American Statistician, 29,* 67.

Selvin, S. (1975b). On the Monty Hall problem. *American Statistician, 29,* 134.

Shaughnessy, J. M. (1992). Research on probability and statistics: Reflections and directions. In D. A. Grouws (Ed.), *Handbook of research on mathematical teaching and learning* (pp. 465–494). New York: Macmillan.

Shepard, R. N. (1987). Evolution of a mesh between principles of the mind and regularities of the world. In J. Dupré (Ed.), *The latest on the best: Essays on evolution and optimality* (pp. 251–275). Cambridge, MA: MIT Press.

Shepard, R. N. (1990). *Mind sights: Original visual illusions.* New York: Freeman.

Shepard, R. N. (1992). The perceptual organization of colors: An adaptation to regularities of the terrestrial world? In J. H. Barkow, L. Cosmides, and J. Tooby (Eds.), *The adapted mind: Evolutionary psychology and the generation of culture* (pp. 495–532). New York: Oxford University Press.

Sherman, L. W. (1992). *Policing domestic violence: Experiments and dilemmas.* New York: The Free Press.

Siegrist, M. (1997). Communicating low risk magnitudes: Incidence rates expressed as frequency versus rates expressed as probability. *Risk Analysis, 17,* 507–510.

Simon, H. A. (1969). *The sciences of the artificial.* Cambridge, MA: MIT Press.

Sjönell, G., & Ståhle, L. (2000). *Mammography screening does not significantly reduce breast cancer mortality in Swedish daily practice* [On-line]. http://www.famnetdoc.com/hhot.htm.

Skolbekken, J.-A. (1998). Communicating the risk reduction achieved by cholesterol reducing drugs. *British Medical Journal, 316,* 1956–1958.

Slaytor, E. K., & Ward, J. E. (1998). How risks of breast cancer and benefits of screening are communicated to women: Analysis of 58 pamphlets. *British Medical Journal, 317,* 263–264.

Slovic, P. (1987). Perception of risk. *Science, 236,* 280–285.

Slovic, P. (1999). Trust, emotion, sex, politics and science. *Risk Analysis, 19,* 689–701.

Slovic, P., & Monahan, J. (1995). Probability, danger, and coercion: A study of risk perception and decision making in mental health law. *Law and Human Behavior, 19,* 49–65.

Slovic, P., Monahan, J., & MacGregor, D. G. (2000). Violence risk assessment and risk communication: The effects of using actual cases, providing instruction, and employing probability versus frequency formats. *Law and Human Behavior, 24,* 271–296.

Spellman, B. A. (1996, October). Degree of difficulty. *American Bar Association Journal, 10.*

Spielberg, F., Kabeya, C. M., Ryder, R. W., Kifuani, N. K., Harris, J., Bender, T. R., Heyward, W. L., & Quinn, T. C. (1989). Field testing and comparative evaluation of rapid, visually read screening assays for antibody to human immunodeficiency virus. *Lancet, 1,* 580–584.

Statistisches Bundesamt (Federal Statistical Office) (Ed.). (2000a). *Statistisches Jahrbuch 2000 für das Ausland (Statistical Yearbook 2000 for foreign countries).* Wiesbaden: Statistisches Bundesamt.

Statistisches Bundesamt (Federal Statistical Office) (Ed.). (2000b). *Statistisches Jahrbuch 2000 für die Bundesrepublik Deutschland (Statistical Yearbook 2000 for the Federal Republic of Germany).* Wiesbaden: Statistisches Bundesamt.

Stein, M. D., Freedberg, K. A., Sullivan, L. M., Savetsky, J., Levenson, S. M., Hingson, R., & Samet, J. H. (1998). Sexual ethics: Disclosure of HIV-positive status to partners. *Archives of Internal Medicine, 158,* 253–257.

Stigler, S. M. (1980). *Stigler's law of eponymy. Transactions of the New York Academy of Sciences, 39,* 147–157.

Stigler, S. M. (1983). Who discovered Bayes's Theorem? *American Statistician, 37,* 290–296.

Stigler, S. M. (1986). *The history of statistics.* Cambridge, MA: Bellknap Press of Harvard University Press.

Stigler, S. M. (1999). *Statistics on the table: The history of statistical concepts and methods.* Cambridge, MA: Harvard University Press.

Stine, G. J. (1996). *Acquired immune deficiency syndrome: Biological, medical, social, and legal issues.* (2nd ed.). Englewood Cliffs, NJ: Prentice Hall.

Stine, G. J. (1999). *AIDS update 1999: An annual overview of acquired immune deficiency syndrome.* Upper Saddle River, NJ: Prentice-Hall.

Svenson, O., Fischhoff, B., & MacGregor, D. (1985). Perceived driving safety and seatbelt usage. *Accident Analysis and Prevention, 17,* 119–133.

Thinès, G., Costall, A., & Butterworth, G. (Eds.). (1991). *Michotte's experimental phenomenology of perception.* Hillsdale, NJ: Erlbaum.

Thompson, W. C. (1993). Worthwhile DNA questions. *Judicature, 77,* 57.

Thompson, W. C., & Schumann, E. L. (1987). Interpretation of statistical evidence in criminal trials: The prosecutor's fallacy and the defense attorney's fallacy. *Law and Human Behavior, 11,* 167–187.

Tierney, J. (1991, July 21). Behind Monty Hall's doors: Puzzle, debate and answer? *The New York Times,* p. A1.

Tjaden, P., & Thoennes, N. (2000). *Full report of the prevalence, incidence, and consequences of violence against women: Findings from the National Violence Against Women Survey.* Washington, DC: U.S. Department of Justice.

Tooby, J. & Cosmides, L. (1992). The psychological foundations of culture. In J. Barkow, L. Cosmides & J. Tooby (Eds.), *The adapted mind: Evolutionary psychology and the generation of culture* (pp. 19–136). New York: Oxford University Press.

Tribe, L. H. (1971). Trial by mathematics: Precision and ritual in the legal process. *Harvard Law Review, 84,* 1329–1393.

Twain, M. (1924). *Mark Twain's Autobiography.* (Vol. 1). New York: Harper and Brothers Publishers.

U.S. Department of Transportation. (1998). *Traffic safety facts 1998* [On-line]. National Center for Statistics and Analysis. Available: http://www.nhtsa.dot.gov.

U.S. Preventive Services Task Force Staff. (1996). *Guide to clinical preventive services: Report of the U.S. preventive services task force.* (2nd ed.). Baltimore, MD: Williams & Wilkins.

vos Savant, M. (1990a, September 9). Ask Marilyn (1). *Parade,* 15.

vos Savant, M. (1990b, December 2). Ask Marilyn (2). *Parade,* 25.

vos Savant, M. (1996). *The power of logical thinking: Easy lessons in the art of reasoning . . . and hard facts about its absence in our lives.* New York: St. Martin's Press.

Ward, J. W. (1994). Testing for human retrovirus infections: Medical indications and ethical considerations. In G. Schochetman & J. R. George (Eds.), *AIDS testing: A comprehensive guide to technical, medical, social, legal, and management issues* (pp. 1–14). New York: Springer.

Was bedeutet Prozent? (1998, December 31). *Süddeutsche Zeitung Magazin,* p. 3.

Windeler, J., & Köbberling, J. (1986). Empirische Untersuchung zur Einschätzung diagnostischer Verfahren am Beispiel des Haemoccult-Tests [An empirical study of the judgments about diagnostic procedures using the example of the Hemoccult test]. *Klinische Wochenschrift, 64,* 1106–1112.

Wingo, P. A., Ries, L. A. G., Rosenberg, H. M., Miller, D. S., & Edwards, B. K. (1998). Cancer incidence and mortality, 1973–1995. *Cancer, 82,* 1197–1207.

Wittkowski, K. (1989). Wann ist ein HIV Test indiziert? Schlusswort. *Deutsches Ärzteblatt, 86,* B-138–140.

Woloshin, S., Schwartz, L. M., Byram, S. J., Sox, H. C., Fischhoff, B., & Welch, G. (2000). Women's understanding of the mammography screening debate. *Archives of Internal Medicine, 160,* 1434–1440.

Wynn, K. (1998). An evolved capacity for number. In D. D. Cummins & C. Allen (Eds.), *The evolution of mind* (pp. 107–126). New York: Oxford University Press.